# Preface

The study of cell population kinetics is a discipline which has matured rapidly in the past decade. Its literature, consequently, is still in the form of key papers, review articles, and symposia, all of which presuppose some knowledge in the mind of the reader. In this little book we attempt to provide that initial knowledge, in some measure at least.

There are three general points about our presentation that the prospective reader should bear in mind throughout his perusal. First, this is not a book of technical recipes. Our aim is to explain the rationale of the procedures by which cell kinetic parameters are worked out. Where it seems to enhance the exposition, technical details will be mentioned. In general, however, the reader must be prepared to go to the references we quote for step-by-step recipes; better still, he should contrive to spend some time in a laboratory where cytokinetic techniques have been established and 'debugged'.

Secondly, the investigation of cell population kinetics, like most forms of biological study, is bedevilled by stochastic error. This is not always openly avowed; one reads about *the* cell birth rate or *the* population doubling time for example, but not always about their *mean* values or their *confidence limits*. There are, indeed, some parameters for which there is no agreed statistical expression. It would be difficult, in a book of this size, to present a rigorous statistical treatment of cell population kinetics, even if we were competent to do so. We shall therefore confine ourselves to occasional 'asides' to keep this source of uncertainty before the reader's mind.

Thirdly, most of the theory we are about to expound is based, necessarily, on simplified models of kinetic phenomena, and not on the real world in its rich diversity. In this we are acting as workers do in most disciplines, especially perhaps biological ones. Inevitably, the reader will detect discrepancies between reality and model, but he should not be scandalized. Better to light a candle than to curse the dark.

We owe debts of gratitude to more people than we can name. We hope it will not seem invidious, therefore, to thank one in particular, Professor A. G. Heppleston, for bringing biologists and pathologists together in the relatively new sphere of cell population kinetics, and for continuous encouragement of our efforts to understand it. A special word of thanks is due to Miss Elizabeth Wark for typing and so patiently retyping such an awkward text. Much of the work embodied in the following pages was supported by grants from the North of England Cancer Research Campaign, to the officers of which we are deeply indebted.

We have tried to eliminate errors and misconceptions from our text but general experience suggests that even the most careful authors do not always succeed completely in doing this. We would be grateful, therefore, to hear from any vigilant reader who discovers errors we have overlooked.

Newcastle upon Tyne
1976

WAA
RSC
NAW

# An Introduction to Cell Population Kinetics

**William A. Aherne,** M.D., Ph.D., F.R.C.Path.
**Richard S. Camplejohn,** M.Sc., Ph.D.
**Nicholas A. Wright,** M.D., Ph.D. M.R.C.Path.

Department of Pathology, University of Newcastle upon Tyne.

Edward Arnold

© W. A. Aherne, R. S. Camplejohn and N. A. Wright, 1977

First published 1977
by Edward Arnold (Publishers) Ltd
25 Hill Street, London W1X 8LL

**British Library Cataloguing in Publication Data**
Aherne, W. A.
   An introduction to cell population kinetics.
   1. Cell populations   2. Cytokinesis
   I. Title   II. Camplejohn, R S   III. Wright, N Λ
   574.8'766          QH587

   ISBN 0-7131–4294–4

In memory of Margaret Aherne

Text set in 10/11 pt IBM Press Roman, printed by photolithography, and bound in Great Britain at The Pitman Press, Bath

# Contents

# The main symbols for cytokinetics

Durations of cell cycle phases:

$t_C$—cell cycle time, i.e., the sum of the phase durations.
$t_{G1}$—duration of the $G_1$ phase.
$t_S$—duration of the S (i.e. DNA-synthesis) phase.
$t_{G2}$—duration of the $G_2$ phase.
$t_M$—duration of mitosis as a whole.
$t_{met}$—duration of metaphase, the major period in mitosis.
$\tau$—cell age as a fraction of the cell cycle.

Intervals related to cell cycle phases or to growth:

$t_{C(a)}$—the apparent cell cycle time (which equals the true cell cycle time when the growth fraction is unity).
$t_P$—occasionally written $t_{Pot}$, the potential population doubling time in the absence of cell loss. Equivalent in duration to $t_{C(a)}$.
$t_D$—the overall population doubling time, in terms of volume ($V$) or mass ($M$).
$t_A$—the duration of metaphase arrest in a stathmokinetic experiment.

Indices:

$I_M$—mitotic index, sometimes written MI.
$I_S$—labelling index after a pulse of isotope (see [3]H-TdR), sometimes written LI.
$I_P$—proliferation index; this is equivalent to the growth fraction ($f_G$).
$I_{met}$—metaphase index (see $t_A$).

Rate constants:

$r_M$—rate at which cells enter mitosis.
$r_S$—rate at which cells enter the phase of DNA synthesis.
$k_B$—cell birth rate.
$k_L$—rate at which cells are lost from the population.
$\phi$ (Steel)—rate of cell loss as a fraction of cell birth, $k_L/k_B$: this may also be written $1 - (t_{C(a)}/t_D)$.
$k_G$—overall growth rate.
$k_{in}$—flux constant into a cell population, or compartment of a population.
$k_{out}$—flux constant out of a population.

SGR——the specific growth rate; explicitly, $M^{-1}(dM/dt)$, i.e. the rate of population mass increase in relation to the mass at the time of observation.

Population compartments or states:

P——the proliferative compartment.

Q——the compartment of non-proliferating cells.

$G_0$——a resting state from which cells may re-enter the division cycle.

E,M (Bresciani)——a compartment in which both expansion (increase in cell number) and maturation occurs.

$d$ (Bresciani)——the ratio $\overleftarrow{\eta}/\overrightarrow{\eta}$ where $\overleftarrow{\eta}$ is the fraction of newly formed cells which continue to replicate and $\overrightarrow{\eta}$ is the fraction which de-cycles and differentiates.

# 1 Introduction

## 1.1 What is cell population kinetics?

The tissues of higher plants and animals are made up of cells and their extra-cellular products. It has been known for over a century that life is maintained by an uninterrupted succession of cells; in other words cells come only from pre-existing cells. A basic property of living organisms and of the cells of which they are composed is the ability to reproduce.

As expected, high rates of cell production are found during embryonic develop-ment and in wound healing or regeneration of some kinds of lost tissue. However, it would be wrong to assume that a static situation exists in the adult organism even though growth has ceased. Life is of necessity a dynamic process and within most tissues of multicellular organisms there is a considerable turnover of the cell population. Cells are constantly produced by division, and lost from a population by death or migration. It has been estimated that the human body loses 1−2 per cent of its cells each day. Therefore if the weight of the body remains constant these cells must be replaced by new ones. This amounts to many millions of cells per day.

The rate of turnover varies enormously between different tissues; for example no new cell production occurs in neurones and little in muscle; on the other hand, turnover is very rapid in such tissues as the small intestinal mucosa and the epidermis. The tissues of other organs such as liver and kidney are intermediate between these two extremes; for example cells are constantly being replaced in the liver but the rate is slow; a liver cell has an average life span of about 18 months compared with 2 days for small intestinal mucosa cells. Consequently, if we look at a section of liver under the microscope we expect to find very few cells in division. However, damage to (or removal of) part of the liver causes a rapid increase in cell production until the original size of the organ is more or less restored. Cell population kinetic studies are aimed at understanding and quantify-ing the dynamic events in the life history of a cell population: these include the processes of proliferation, differentiation, migration, and death.

## 1.2 Cell division

The phenomenon of mitosis has long been known. Detailed descriptions are readily available of the sequential events which comprise it. Mitosis is the normal pattern of nuclear division found throughout the animal and plant kingdoms. It achieves the separation of identical genomes into the two daughter cells produced. In animal cells it is usually closely linked in time to the process of cell division. In plants the delay between completion of nuclear and cell division may be greater, due, presumably, to the need to synthesize a new cell wall between the daughter cells. In certain conditions, usually pathological or experimental, the synchrony between nuclear and cell division may break down, resulting in the production of multinucleate cells. Kinetic techniques in general relate to *nuclear* rather than *total cellular* division.

Though mitosis has been known for many years it is only recently that the existence of a *division cycle* has been recognized. The division cycle which proliferating cells traverse is an ordered sequence of biochemical events leading up to mitosis. Mitosis is the only visually distinct phase but, as we shall see, other phases can be made visible by techniques now available.

## 1.3  The control of cell division

It is still far from clear *why* cells grow and divide, or conversely, why cells stop dividing at the appropriate time. In unicellular organisms, such as amoebae, there is little doubt that cell size is an important factor. Division would seem to be an attempt to maintain a critical relationship between surface area and mass. This is presumably related to the ability of a cell to obtain a sufficient supply of substances required for its metabolism. Surface area increases only as the two-thirds power of mass; thus if the mass of the cell becomes too large its centre may become deficient in metabolites such as oxygen.

Rashevsky (1961) has shown, by an approximate but general biophysical deduction, that the average concentration [c] of a given substance in a *spherical* cell may be written

$$[c] = [c_0]/[1 - (kr^2/6D)] \; ; \; r < 6D/k; \tag{1.1}$$

where [c_0] is the constant external concentration of the substance in question, *r* is the radius of the cell, *D* is the appropriate diffusion coefficient and k is a reaction rate constant. It is clear from this equation that an increase in the radius of the cell will lead to an increase in [c] and so, possibly, to a metabolic imbalance. Whatever the reason, division into daughter cells maintains a surface/volume relationship which is characteristic of the cell in question.

## 1.4  The control of population size

In multicellular organisms many other factors are likely to be involved in control processes. Despite research and debate over many years, notably since the 1950s, there are still few solid facts. The following paragraphs are intended as a rough sketch of this complex territory.

Goss (1972) epitomizes the matter as follows. 'Existing hypotheses of growth regulation can be categorized in several ways. Some propose the operation of inhibitors; others propose stimulators. Some would have each organ and tissue control its own growth. Others put the regulator elsewhere in the body . . .' He then isolates two main schools of thought. 'One contends that the dimensions of body parts are genetically predetermined. The other holds that the correct size of an organ is a function of the physiological demands impinging on it.' He then proceeds to balance the pros and cons of the various major hypotheses.

### 1.4.1 *Control by self-inhibition*

A popular hypothesis is that each organ or tissue regulates its own growth and appropriate mass by a *negative feedback mechanism*. Weiss and Kavanau (1957) suggested a model in which 'templates' normally stimulate synthetic processes and cell growth, but may be appropriately inhibited from doing so by 'anti-templates'. A growing organ, in this hypothesis, produces an increasing number of antitemplates into the body space until their concentration is sufficient to inhibit further growth. Loss of tissue would reduce the ambient concentration of antitemplates and so lead to renewed growth.

A similar concept, with a slightly different mechanism, is that maturing cells of a given population secrete a substance which diffuses into the proliferative (stem cell) compartment and regulates further growth by an appropriate degree of inhibition. This idea is particularly associated with the names of Bullough and Laurence in Great Britain, and Rytomaa and Kiviniemi in Finland. Both schools refer to their inhibitors as *chalones*; epidermal in the former and granulocytic (leucocytic) in the latter. Recent work by Thornley and Laurence (1975) suggests that the epidermis secretes two different chalones, one inhibiting the flux of proliferating cells into the phase of DNA synthesis, and the other inhibiting entry into mitosis (Fig. 1.1). It is particularly interesting that chalones (whose chemical nature has not yet been established to everyone's satisfaction) are tissue-specific but not species-specific.

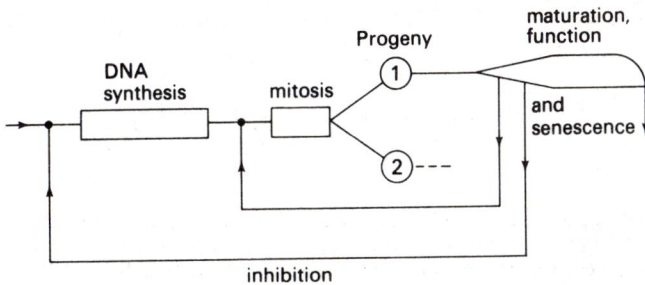

**Fig. 1.1** This diagram illustrates the hypothesis that cell proliferation may be regulated by negative feedback inhibition. According to the 'chalone' concept at least one inhibitor is produced by maturing cells which can inhibit precursors of the same type from entering mitosis (the $G_2$ – chalone; q.v.), and at least another (the $G_1$ – chalone) inhibits precursor cells, again of the same type, from embarking on DNA synthesis.

### 1.4.2 *Control by contact inhibition*

Chalones have been shown to inhibit cell proliferation *in vivo* and *in vitro*. But perhaps the most conclusive *in vitro* experiments concern a rather different mode of growth control, namely *contact* inhibition. Normal fibroblasts grow in tissue culture until a confluent monolayer is formed. In this state of extensive mutual contact both growth and movement are inhibited. The work of Burger in particular indicates that this form of regulation is dependent upon the structure of the cell membrane. For instance, a brief exposure of normal 3T3 fibroblasts to trypsin and other proteases confers proliferative properties, *in vitro* and *in vivo*, similar to those of cells which have been transformed to cancer cells (i.e. 'malignant' cells) by certain viruses. This is a reversible phenomenon; trypsin-treated cells appear to be capable of regenerating a normal plasma membrane on subculture in the absence of the enzyme. There is sound evidence that trypsin removes an outer coat and thus exposes sites on the cell membrane which can be artefactually detected by plant agglutinins such as concanavalin A (con-A). Cancer cells are similarly agglutinable. Burger and Noonan (1970) showed that these sites can be covered again by *monovalent* con-A, which has only a slight ability to agglutinate. The intriguing outcome of this artificial covering is that the cells in question cease to behave as cancer cells and recover the normal property of contact inhibition. Figure 1.2 illustrates this experiment in diagrammatic form.

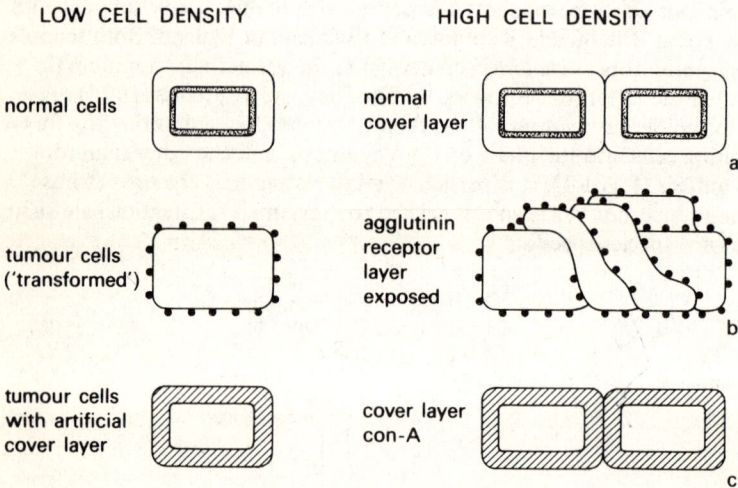

**Fig. 1.2** Diagram illustrating (a) cells with normal outer membrane undergoing contact inhibition; (b) cells after treatment by trypsin, with an external layer of their membranes removed and agglutination sites exposed, no longer capable of contact growth inhibition; and (c) cells artificially reclothed in monovalent concanavalin A, with agglutination sites covered and contact inhibition restored (redrawn from Burger and Noonan, 1970).

### 1.4.3 *Control by stimulation, exogenous or endogenous*

The notion that cell numbers, or tissue mass, may be regulated primarily by *stimulation*, exogenous or endogenous, has found less favour than negative feedback self-inhibition. Burwell (1963) developed a complex hypothesis of an intermediate kind. He proposed that proliferating tissues release *tissue-specific factors* which, we may suppose, behave rather like Bullough's chalones. But he assigned to lymphatic tissue a central role in monitoring and regulating the proliferation of other tissues by means of humoral and cell-bound globulins of appropriate clones. Partial removal of an organ, for example, is supposed in this hypothesis to induce an increased output of cells bearing globulins complementary to the specific factors of that organ. The concentration of specific factors would thus be reduced. This, and the secretion of *mitotic control proteins*, also by lymphatic tissue, would lead to mitotic activity and so to a restoration of the original mass.

The most characteristic feature of mammalian growth, as Tanner's (1963) studies of *human* growth have shown, is that it is self-stabilizing. Tanner speculated that the brain of the growing mammal might keep a tally of normal growth, and of whether or not there was any mis-match between optimal size and actual size at a given time. The tally, he supposed, might take the form of a steady increase in neuronal concentration of some time-related substance. He postulated that a second substance, complementary to the first, accumulates at a rate proportional to protein synthesis during growth; a mis-match *might* then be represented as a deficiency of the second substance relative to the first, and this *might* act as a signal for the secretion of a growth stimulator. The concentration of the stimulator would, in this hypothesis, be proportional to (or a function of) the mis-match, and the velocity of catch-up in turn would be proportional to (or again a function of) the concentration of stimulator.

Hypotheses of exogenous control may be plausible and reasonable abstractions from indisputable observations, but they cannot explain the way in which the exogenous master-tissue is itself regulated.

### 1.4.4 *Cyclic AMP and cyclic GMP*

We saw, in Section 1.4.2 how certain kinds of cell behaviour appear to be regulated by reactions taking place at cell surfaces. The control of cell movement and, in particular, of cell proliferation requires the presence of an intracellular mediator between the plasma membrane and various internal structures. Insulin, for example, exerts its effect upon a cell without entering it; it reacts with a surface receptor and in due course a number of biochemical processes are triggered off.

There is now much evidence that the mediator between the cell's membrane and its interior is *cyclic adenosine monophosphate*, or cyclic AMP. Cyclic AMP is involved in many reactions; its relevance to cell population kinetics is clear from the following observations. The concentration of cyclic AMP in rapidly growing cells is low. When a confluent monolayer of cells is formed in tissue culture the concentration of cyclic AMP rises significantly. If at this stage fresh serum or insulin is added, proliferation begins again, after the usual synthesis of RNA and protein followed by synthesis of DNA. These reactions are accompanied by a fall in the concentration of cyclic AMP. If cyclic AMP is added to the medium before the addition of serum or insulin, the reactions leading to cell proliferation are inhibited. Cells which have been transformed by oncogenic (i.e. tumour inducing) viruses have very much lower levels of cyclic AMP than normal fibroblasts, even at confluence. Addition of cyclic AMP to tumour cells in culture inhibits their proliferation. Indeed, the addition of cyclic AMP to cells derived from a tumour of neural origin is followed by a cessation of DNA synthesis, the formation of dendrites and the secretion of neurotransmitters; in other words, these cells have apparently switched from proliferation to differentiation.

More recently a second cyclic nucleotide, *guanosine cyclic monophosphate* or cyclic GMP, has been found to occur at concentrations inverse to cyclic AMP. For example, the addition of insulin to confluent, static cell cultures is followed by a rise in the level of cyclic GMP as well as a fall in cyclic AMP. These and other experiments have led to the hypothesis that high levels of cyclic GMP and low levels of cyclic AMP set the stage for cell proliferation and, conversely, that low levels of cyclic GMP and high levels of cyclic AMP induce cells to leave the proliferative cycle and differentiate. This hypothesis of dual control has been fancifully likened by Goldberg (quoted by Hogan and Shields, 1974) to the Oriental philosophy of opposed but complementary forces, Yin and Yang. Clearly, within this hypothesis there is a further hypothesis about the way in which a cell might be released from its normal restraints in neoplastic transformation.

If this brief sketch of population control theory is at least roughly representative then there are many problems and relatively few firm solutions in this area of biology. No doubt solutions will be reached in due course, and the rich mathematical theory of systems control may well provide a ready-made language. We do not have to wait upon such solutions; in cell population kinetics we are concerned primarily with *measuring rates*, e.g. of cell progress through the mitotic cycle, rather than with elucidating how such phenomena are controlled. Should the reader wish to pursue topics in cell population control we recommend the monograph by Goss (1972) and that published on chalone research by the National Cancer Institute in 1973. Both are listed at the end of this chapter. The chapters that now follow deal with the commoner techniques by which cytokinetic rate parameters and state parameters are calculated.

## Summary

Cell population kinetics is a body of methods for quantifying certain state parameters and certain rate parameters by which a proliferating mass of cells may be characterized.

The mechanisms by which cell division and population size are controlled are imperfectly understood. This area of biology, though deeply interesting, is not primarily a part of cell population kinetics.

## References

BULLOUGH, W. S. and LAURENCE, Edna (1968a) Epidermal chalone and mitotic control in the V x 2 epidermal tumour. *Nature (Lond.)*, 220, 134.

BULLOUGH, W. S. and LAURENCE, Edna (1968b) Melanocyte chalone and mitotic control in melanomata. *Nature (Lond.)*, 220, 137.

BURGER, M. M. and NOONAN, K. D. (1970) Restoration of normal growth by covering of agglutinin sites on tumour cell surface. *Nature (Lond.)*, 228, 512.

BURWELL, R. G. (1963) The role of lymphoid tissue in morphostasis. *Lancet*, ii, 69.

GOSS, R. J. (Editor) (1972) *Regulation of Organ and Tissue Growth*, p. 1. Academic Press, New York and London.

HOGAN, Brigid and SHIELDS, R. (1974) Yin-Yan hypothesis of growth control. *New Scientist*, May 9th, p. 323.

RASHEVSKY, N. (1961) *Mathematical Principles in Biology*, p. 11. Charles C Thomas, Springfield, Ill.

RYTOMAA, T. and KIVINIEMI, K. (1968) Control of cell production in rat chloroleukaemia by means of the granulocytic chalone. *Nature (Lond.)*, 220, 136.

TANNER, J. M. (1963) Regulation of growth size in mammals. *Nature (Lond.)*, **199**, 845.

THORNLEY, A. L. and LAURENCE, Edna (1975) Chalone regulation of the epidermal cell cycle. *Experientia*, **31**, 1024.

WEISS, P. and KAVANAU, J. L. (1957) A model of growth and growth control in mathematical terms. *J. Gen. Physiol.*, **41**, 1.

## Suggestions for further reading

BASERGA, R. (Editor) (1971) *The Cell Cycle and Cancer*. Marcel Dekker Inc., New York.

BUSCH, H. (Editor) (1971) Studies on tumor cell population kinetics. In *Methods in Cancer Research*, Vol. 6, p. 4. Academic Press, New York.

CLEAVER, J. E. (1967) *Thymidine Metabolism and Cell Kinetics*. North Holland, Amsterdam.

GOSS, R. J. (Editor) (1972) *Regulation of Organ and Tissue Growth*. Academic Press, New York.

LALA, P. K. (1972) Age-specific changes in the proliferation of Ehrlich ascites tumor cells grown as solid tumors. *Cancer Res.*, **32**, 628.

MACHIN, D. (1976) *Biomathematics: An Introduction*. The MacMillan Press Ltd.

MITCHISON, J. M. (1971) *The Biology of the Cell Cycle*. Cambridge University Press.

NATIONAL CANCER INSTITUTE MONOGRAPH 38 (1973) *Chalones: Concepts and Current Researches*.

SNEDECOR, G. W. and COCHRAN, W. G. (1971) *Statistical Methods, 6th Edition*. The Iowa State University Press, Ames, Iowa.

SYMINGTON, T. and CARTER, R. L. (Editors) (1976) *Scientific Foundations of Oncology*. William Heinemann Medical Books Ltd, London.

# 2 Cell cycle events

An essential aspect of cell proliferation in most cases is that each daughter cell should have exactly the same genome as the mother cell. Therefore, the quantity and profile of the mother cell's DNA must be replicated before the cell divides. We may note, in passing, that the replication of DNA is *semi-conservative*; that is, half of the DNA in each daughter cell is newly synthesized and half is carried over from the previous generation. In 1953 it was discovered, by following the uptake of radioactively labelled precursors, that DNA was synthesized during a *limited* part of interphase and not during mitosis as had been supposed. This led to the concept of the *cell cycle*, in which the cell goes through a number of distinguishable phases between one mitosis and the next. The period of DNA synthesis, during which, in mammalian cells, some 250 000 adenine–thymine and guanine–cytosine base pairs are assembled *per second* (Lajtha, 1970), occupies a slightly variable position in the latter half of the cycle.

## 2.1 Distinguishable phases of the cell cycle

The period of DNA synthesis is commonly designated the *S phase*. The interval between the previous mitosis and the S phase, during which the cell may manifest its characteristic function, is called the $G_1$ *phase*, and the interval following S and leading into the next mitosis is called the $G_2$ *phase*. The symbols $G_1$ and $G_2$, merely stand for 'gaps' in the proliferative cycle. The use of such non-specific symbols should not conceal the fact that cellular events related specifically to the cycle must take place during these periods. There is a good deal of evidence, for example, to suggest that the late $G_1$ phase is an important 'decision point' in the cycle. It is suggested that at this time cells either (a) irreversibly embark on the sequence of metabolic events leading to mitosis, or (b) *decycle*, i.e. cease to pass around the cycle and enter a resting phase. Whether this is the only control point in the cycle is not certain but it is clear that essential metabolic events must occur within the cell in late $G_1$ before DNA synthesis can begin. Experiments with antibiotics such as puromycin and actinomycin D have shown that the synthesis of both RNA and protein are necessary for progress through the $G_1$ phase and entry into S. Similarly RNA and protein synthesis are necessary for cells to pass through $G_2$ and enter mitosis. At least part of this protein synthesis is likely to involve the induction of enzymes and initiators involved in cell proliferation. A good deal of research is currently afoot in this area.

Figure 2.1 shows the cell cycle and its four phases, in roughly typical propor-

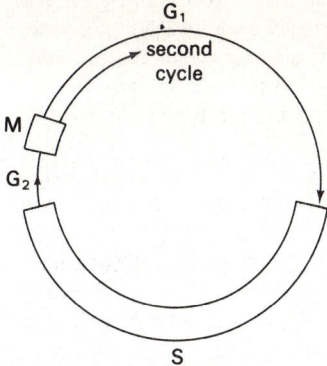

**Fig. 2.1** The cell division cycle in its simplest form. Gap 1 (G$_1$) is the most variable phase; occasionally it is undetectable.

tions of the cycle time. G$_1$ is the phase which varies most in duration but none are inflexible. In the simplest possible kinetic situation (rarely if ever found in mammalian studies *in vivo*, though it can be approximated *in vitro*), every cell in the population traverses the cycle and no cells are lost. The time for a cell to go once around the cycle is called the *cell cycle time* and is commonly denoted $t_C$. In the situation shown in Fig. 2.1 the population doubles its number of individuals in a period of time equal to $t_C$. In real populations not all cells have exactly the same cycle time. The value of $t_C$ quoted for a population will be an average, and an ill-defined average at that, since it is difficult to enclose it within confidence limits. The average $t_C$ itself varies from one cell population to another.

## 2.2 Population states associated with the cell cycle

Real kinetic situations are more complicated than depicted in Fig. 2.1. In most real cell populations not all of the cells are in the proliferative cycle; the population is a mixture of proliferating and non-proliferating cells. This fact leads to the important concept of the *growth fraction*, which may be defined, in *index form*, as

$$I_P = N_C/N \qquad (2.1)$$

where $N_C$ stands for the number of proliferating cells, $N$ for the total number of cells in the population, and $I_P$ is the proliferative index. Some authors use the literal contractions GF or $f_G$ for 'growth fraction'; the symbol $I_P$, though perhaps less immediately meaningful, conforms with the other indices of population kinetics, as we shall see.

If some of the cells are not proliferating then it follows that the rate of cell production in the population as a whole will be slower than expected from the value of $t_C$. The time taken for the number of cells to double (assuming no cell loss) will not equal $t_C$ as it did in the situation portrayed in Fig. 2.1; it will take longer than $t_C$. This time period has been called the *potential population doubling time* ($t_P$) and may also be referred to as the *apparent cell cycle time* ($t_{C(a)}$). Note that we have introduced two words, 'proliferation' and 'production', which are sometimes used to convey slightly different kinds of information. Proliferation may be taken to denote the birth process whereby a cell in cycle produces two daughter cells. Production, on the other hand, is a more general word which is used in discussing the output of new cells in relation to the population as a whole. They are not mutually exclusive words: proliferation is a component of production.

The rate of cell production is determined jointly by the size of the growth fraction and by the cycle time of the cells which comprise that fraction. In some circumstances, for example the more indolent of the human malignant tumours, the growth fraction may be as small as 0·05—0·1 (5—10 per cent). In such instances the cell production rate of the population will be very low, even though the quantum of proliferating cells may be very active, with short cell cycle times. It should be noted also that the rate of cell loss is as important as the rate of cell production in deciding the size of a cell population. The state of balance between these two last rates determines whether a cell population will remain static, expand (as in neoplasms), or decline.

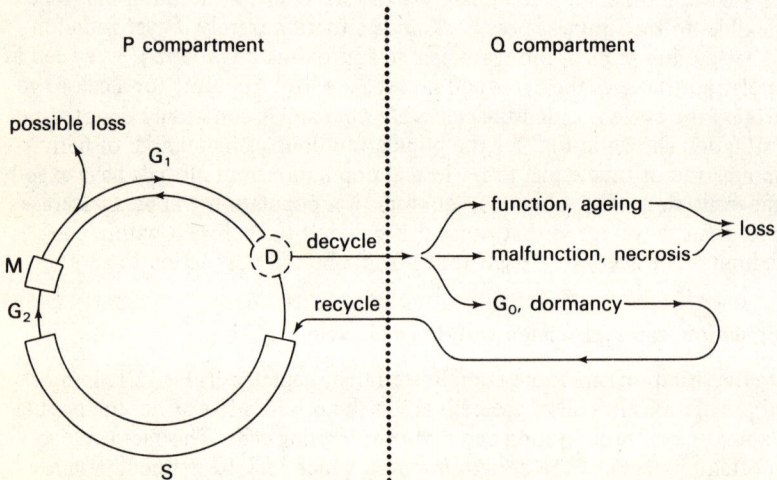

**Fig. 2.2** A more complex form of the cell division cycle showing how the population is partitioned into proliferating cells and non-proliferating cells, and how cells in the resting $G_0$ phase may return to the division cycle.

Figure 2.2 shows a somewhat more complicated kinetic model into which we have introduced a *non-cycling* fraction of cells. It also illustrates that cells which leave the cycle may (a) differentiate irreversibly, function, and finally die effete, (b) undergo immediate necrosis, or (c) enter a resting state from which they may return to cycle. The last state (c) has been termed $G_0$. Cells in the $G_0$ stage are fertile and retain an ability to enter the active cell cycle under proper conditions or stimuli. Most liver cells are normally in a $G_0$ state and will re-enter the cycle if part of the liver is removed. By this means the lost liver tissue is replaced. Salivary gland cells in $G_0$ may be stimulated to proliferate if they are treated with isoproterenol. In cases such as this the $G_0$ cells undergo a series of metabolic events preceding DNA synthesis. These are illustrated in Fig. 2.3.

| OH group | protein synthesis | unknown RNA templates | | UMP kinase | protein synthesis | glycogen synthesis | | TMP synthetase DNA polymerase | | DNA synthesis | |
| secretion | | | | | | templates for TdR and TMP kinases, TMP synthetase | glycogen breakdown | Thymidine and TMP kinases | | | mitosis |

```
0        5        10       15       20       25       30   33
↑
IPR
```

time (hours)

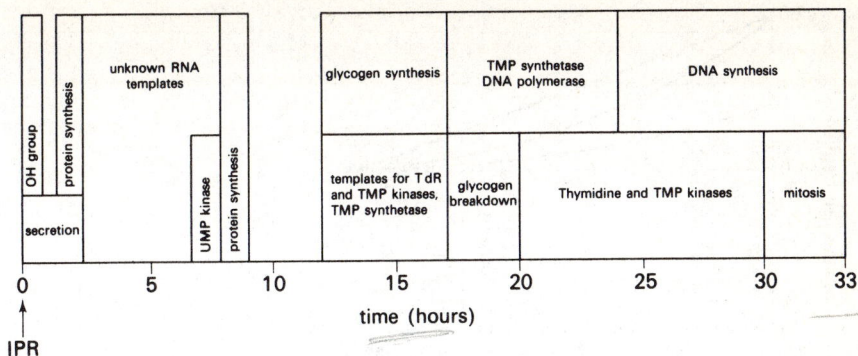

**Fig. 2.3** This diagram shows some of the biochemical events, prior to mitosis, known to occur in murine or rat salivary glands after a single injection of isoproterenol. The diagram is drawn from the work of Fujioka, Koga and Lieberman (1963); further comment may be found in Baserga (1971).

## 2.3 Age distribution of cells within a population

Most cell populations are not synchronized but contain cells whose age points are distributed throughout the cell cycle. It is possible experimentally to obtain populations with quite a high degree of synchrony. We shall not be concerned with such populations. Indeed, most kinetic techniques are *based on the assumption* that cells proliferate asynchronously.

Let us define cell age as a fraction $\tau$ of the whole cycle with $t_C = 1$. The *frequency*, in the population, of cells with a given value of $\tau$ depends on whether the population is (a) in a steady state (e.g. epidermis considered over all), or (b) expanding (e.g. an embryonic organ or a cancer). In the steady state the frequency ($n_t$) of cells of a given age $\tau_i$ hours is uniform. In expanding populations (if we make the common simplifying assumption that growth is exponential), the frequency of cells of a given age has an exponential distribution of the form

$$n_t = f(\tau) = n_0 \exp\left[(-\ln 2)\,\tau/t_C\right] \tag{2.2}$$

which formally expresses the intuitive idea that older cells are fewer than younger cells. In the limiting case young cells, having just emerged from mitosis, are twice as frequent in the population as old cells (as $\tau \to t_C$).

The age distribution for the exponential case, idealized, of course, is represented by line 2 in Fig. 2.4. In steady state populations, on the other hand, cell production equals cell loss and there is no change in population size with time. In an idealized steady state system, where cell loss is assumed to occur randomly with respect to the phases of the cell cycle, there is a rectangular age distribution of cells over the cycle (the shaded area in Fig. 2.4). In such a case the number of cells in any phase of the cycle is proportional to the duration of that phase. Thus

$$I_M = t_M/t_C \tag{2.3}$$

and

$$I_S = t_S/t_C \tag{2.4}$$

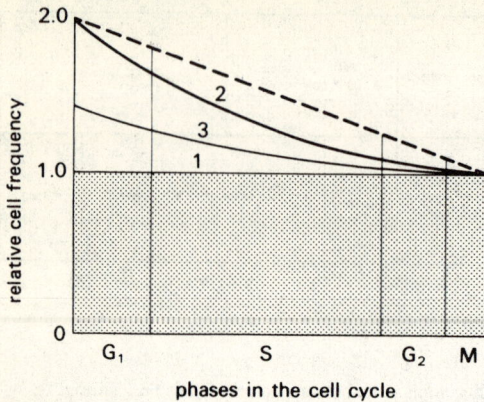

**Fig. 2.4** Age distribution of cells along the cell cycle for two model populations. The shaded rectangle (1) represents the expected age distribution in a steady state population. The areas bounded by the curved lines (2 and 3) above and the abscissa below represent the age distribution in expanding populations; in the case of curve 2 every cell in the population is replicating, and therefore the population doubles each cycle; in the case of curve 3 only a proportion of cells are actively replicating, and therefore the population does not double. For some computations the uppermost (dashed) line may be used as an approximation. Redrawn from Lala, 1971.

where $I_M$ is the *average mitotic index* or the fraction of cells in mitosis at a point in time, $t_M$ is the *average duration of mitosis*, and $t_C$ is the average cell cycle time. $I_S$ is the proportion of cells in the S phase, i.e. the *labelling index* (see Chapter 4). Some authors have used the literal contractions MI and LI for the mitotic index and the labelling index respectively.

### 2.3.1 *The effect of age distribution on kinetic calculations*

In an ideal, exponentially growing population, where there is no cell loss, no synchrony, and all cells are proliferating the number of cells doubles in a period equal to $t_C$. It is evident from Fig. 2.4 (line 2) that twice as many cells will enter $G_1$ as enter mitosis at any given instant, and the cell flux from one phase to the next decreases with the distance away from mitosis in the cycle. Consequently the *average rate of entry* of cells into DNA synthesis ($r_S$) is greater than the *average rate of entry of cells into mitosis* ($r_M$). The fraction of the population in any given phase of the cycle is equivalent to the ratio of the area under the (exponential) boundary curve occupied by cells in that phase to the total area under the curve. These fractions can be measured graphically if the age distribution diagram is available. Alternatively, relationships for average mitotic and average labelling indices may be obtained from the following equations derived from the age distribution:

$$I_M = \ln 2(t_M/t_C) \tag{2.5}$$

and

$$I_S = [\exp(t_S \ln 2/t_C) - 1] [\exp t_2 \ln 2/t_C]. \tag{2.6}$$

These equations taken together with **2.3** and **2.4** illustrate how calculations of kinetic parameters depend upon the age distribution of the cell population. Strictly, they are exact only if all cells in the population are proliferating. This is often not the case. Some cells move into $G_0$, some differentiate, and some die. The vertical ordinate of the age distribution diagram at the beginning of $G_1$ is thus reduced (line 3 on Fig. 2.4). It follows that for every 10 cells entering mitosis in these circumstances fewer than 20 enter $G_1$.

If cell loss were to occur randomly from *both* proliferative (P) *and* quiescent (Q) compartments of the population, the shape of the age distribution diagram would not be disturbed. If, as is likely in real populations, loss occurs selectively, our usual assumptions become less valid and error creeps into our calculations. In the absence of information to the contrary, it is practically impossible to avoid assuming that cell loss occurs randomly. But it is important to be fully aware of the assumptions we are making and of their implications.

### 2.3.2 *Decycling and recycling*
If a proportion of cells opt out of the proliferative cycle after mitosis, the consequent reduction in the vertical ordinate of the age distribution may be expressed in terms of the growth fraction as

$$\alpha = (I_P + 1) < 2. \tag{2.7}$$

It follows that the probability of their leaving the cycle (i.e. of 'decycling') is

$$2 - \alpha = 1 - I_P \tag{2.8}$$

Bresciani (1968) proposed a different notation for these phenomena. He represented the number of cells which decycle as $\vec{\eta}$ and the number which recycle as $\overleftarrow{\eta}$. He termed the ratio $\overleftarrow{\eta}/\vec{\eta}$ the 'distribution ratio' $(d)$, and commented that progressive growth can be maintained only if the ratio $\overleftarrow{\eta}/\vec{\eta}$ is greater than unity. In practice it is more convenient to work in terms of the decycling probability, and we shall see how it may be calculated presently (**4.8**, p. 39; but see **5.5**, p. 50 also).

Expanding populations do not necessarily entail the use of exponential formulae. When the growth fraction is small and the cell loss rate is large, the height of the age distribution curve approximates to the steady state level. Indeed, the age distribution curve of cells within an expanding population can lie anywhere between the two limit curves of exponential growth and steady state, respectively.

## 2.4 Cell birth and loss rates and population growth rates

The *average birth rate* ($k_B$) of new cells (the production rate) is jointly determined by the growth fraction and the average cell cycle time ($t_C$). Thus for: *steady state populations*

$$k_B = I_P/t_C = r_M \tag{2.9}$$

*exponentially growing populations*

$$k_B = \ln (1 + I_P)/t_C = r_M. \tag{2.10}$$

Cell birth rate ($k_B$) is expressed as *new cells per cell per hour*. The time taken for a cell population to double in number, i.e. the apparent cell cycle time $(t_{C(a)})$, is related quite simply to the birth rate thus:

*steady state*

$$k_B = 1/t_{C(a)} \qquad\qquad (2.11)$$

*exponential growth*

$$k_B = \ln 2/t_{C(a)} \qquad\qquad (2.12)$$

In the absence of cell loss the birth rate is equal to the growth rate $(k_G)$ of the population. The *growth rate*, $k_G$, is the actual net rate of increase in the number of cells in a population, and like $k_B$ is expressed as new cells per cell per hour. If cell loss is occurring, $k_G$ will be less than $k_B$ and the discrepancy between the two is a measure of the rate of cell loss $(k_L)$:

$$k_G = k_B - k_L. \qquad\qquad (2.13)$$

A parameter often used, particularly in tumour cell kinetics, is the average overall *population doubling time* $(t_D)$. In an exponentially growing population this is related to the growth rate thus:

$$t_D = \ln 2/k_G. \qquad\qquad (2.14)$$

Cell loss is often expressed as a fraction or percentage of total cell birth per unit of time. It is symbolized as $\phi$ (Steel, 1968). This quantity denotes the magnitude of the loss rate in relation to the birth rate, $k_B$. The cell loss factor can be calculated if we know $k_L$ and $k_B$ or, of course, if we know $k_G$ and $k_B$. As $t_{C(a)}$ and $t_D$ are related simply to $k_B$ and $k_G$, we can also use these parameters to estimate the cell loss factor.

$$\phi = k_L/k_B = 1 - (t_{C(a)}/t_D). \qquad\qquad (2.15)$$

The relationship between cell cycle time, apparent cell cycle time, doubling time, growth fraction, and cell loss are summarized in Fig. 2.5.

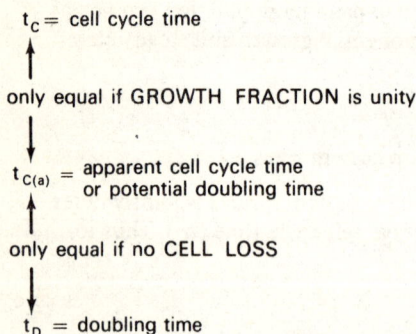

$t_c$ = cell cycle time

$\uparrow$

only equal if GROWTH FRACTION is unity

$\downarrow$

$t_{C(a)}$ = apparent cell cycle time
or potential doubling time

$\uparrow$

only equal if no CELL LOSS

$\downarrow$

$t_D$ = doubling time

**Fig. 2.5** A scheme showing the relationships of the true cell cycle time $(t_C)$, the apparent cell cycle time $(t_{C(a)})$ which is longer than $t_C$ because the population contains non-proliferating cells, and the overall volume doubling time $t_D$, which is the longest measure due to the further factor of cell loss.

## 2.5 How these equations are used in practice

Equations 2.3 to 2.6 are stated above in forms which evolve naturally from their derivations. Thus, for example, $I_M = t_M/t_C$ is an expression of an idealized situation where we regard the cell cycle time, $t_C$, as a line segment, and the duration of mitosis, $t_M$, as a small subsegment of that line. Clearly, the frequency of mitosis can be represented as the fraction $t_M/t_C$. But in practice we are more likely to want to know the value of $t_C$ (given $t_M$ and $I_M$), and we shall write Equation 2.3 in the form $t_C = t_M/I_M$. The value of $I_M$ does not, in fact, need to be calculated; it is an *observable* parameter, established by counting mitoses and expressing their number as a proportion of all the cells in the population. Equation 2.4, similarly, enables us to calculate the cell cycle time when it is cast in the form $t_C = t_S/I_S$. Again $I_S$ is always an observable parameter, and we may know $t_S$ from another experiment (see Chapter 4).

Equation 2.6 is not often used to calculate $t_C$; in fact, it cannot be rewritten explicitly to solve for $t_C$. It is an equation which must be tackled by means of an iterative process such as the Newton–Raphson method of successive approximations.

The cell birth rate, $k_B$ is defined as $I_p/t_C$ (or $1/t_{C(a)}$) and is therefore equivalent to the rate of entry into mitosis, $r_M$. The value of $r_M$ is an observable quantity, as we shall see in Chapter 3 (Equations 3.1 and 3.2). Equation 2.9 may be used to calculate $t_C$ if both $r_M$ and the growth fraction $I_p$ are known from other experiments. Similar comments apply to Equation 2.10.

Equation 2.14 is more likely to be used in the form $k_G = \ln 2/t_D$ since the overall volume doubling time, $t_D$, is again a measured quantity. If all of the cells in the population are in cycle, and there is no cell loss, then $k_G = k_B$; but this is not a realistic situation and in practice $k_G$ must be calculated from measurements of $t_D$.

## Summary

There are four phases in the proliferative cell cycle: (1) a phase of function, concluded by preparations for; (2) a phase of DNA replication; then (3) a phase concluded by preparation for mitosis; and finally (4) mitosis. In (2) cells can be labelled by radioactive thymidine. In (4) they can be arrested by certain drugs (see Chapter 3).

Usually, only a proportion of cells are in the proliferative cycle. This proportion constitutes the growth fraction, $I_p$ (or $f_G$). The remainder of the population is in a resting state ($G_0$), or is irreversibly differentiated, or is dying.

In a steady state, where additions to the population are balanced by subtractions from it, the age distribution is rectangular. This gives rise to particularly simple equations for the mitotic index and the labelling index, $I_M = t_M/t_C$ and $I_S = t_S/t_C$, respectively, since the fraction of the population in any particular phase is proportional to the duration of that phase. These equations are more likely to be used in the form $t_C = t_M/I_M = t_S/I_S$.

In a state of exponential growth, the equations for $I_M$ and $I_S$ must take into account the fact that there is a greater proportion of young cells than of old cells in the population. The age distribution curve takes the form of a negative exponential, and accordingly $I_M = \ln 2(t_M/t_C)$, and therefore $t_C = \ln 2(t_M/I_M)$.

Since $k_B$, the cell birth rate, measures the number of new cells formed per unit number of cells per hour, it may be conceived alternatively as the rate of entry into mitosis $r_M = I_P/t_C$, or $1/t_{C(a)}$.

In real populations, where cells are continually being lost, the over-all cell production rate is the difference between birth rate and death rate: $k_G = k_B - k_L$. The rate $k_G$ is related to the measurable over-all volume doubling time: $t_D = \ln 2/k_G$. Cell loss may be expressed as the factor $\phi = k_L/k_B = 1 - (t_{C(a)}/t_D)$; this concept relates the rate of loss to the rate of production.

## References

BASERGA, R. (Editor) (1971) *The Cell Cycle and Cancer*, p. 191. Marcel Dekker Inc., New York.

BRESCIANI, F. (1968) Cell proliferation in cancer. *Europ. J. Cancer*, 4, 343.

FUJIOKA, M., KOGA, M. and LIEBERMAN, I. (1963) Metabolism of ribonucleic acid after partial hepatectomy. *J. Biol. Chem.*, 238, 3401.

LAJTHA, L. G. (1970) The nature of cancer. In *What We Know About Cancer*, ed. Harris, R. J. C., p. 34. George Allen and Unwin Ltd., London.

LALA, P. K. (1971) Studies in tumour cell proliferation kinetics. In *Methods in Cancer Research*, vol. 6, 3. Academic Press, New York.

STEEL, G. G. (1968) Cell loss from experimental tumours. *Cell Tiss. Kinet.*, 1, 193.

# 3 How to analyse the cell cycle: I

## Techniques based on counting mitoses

In chapter 2 we described the basic cycle which proliferating cells traverse and we defined the major cytokinetic parameters. In this and the next two chapters we will outline some of the more important techniques by which these parameters may be estimated. Though cell population kinetics is a relatively new subject, advances in methodology have been rapid. Our account of the currently popular techniques cannot therefore be fully comprehensive.

Most cytokinetic methods depend upon conventional microscopical sections, on which, for example, counts of metaphases, $^3$H-labelled cells, and $^3$H-labelled metaphases are made. Two factors are essential in the preparation of good sections: namely, gentle handling of unfixed tissue, and adequate fixation before the tissue is embedded, sectioned, and stained.

### 3.1 The sample cell count for the mitotic index

The question immediately arises as to how many cells we must count to obtain a reliable estimate of the mitotic index. Of course, there is the prior matter of obtaining a satisfactory sample on which the count will depend for its *accuracy*, i.e. the proximity of its mean to the population mean, which, by the way, does not necessarily have a bearing on the *precision* of the count, i.e. its standard error. The best procedure is often some form of *stratified* random sampling. For further guidance the reader should consult one of the recommended books, or discuss the matter with a statistician. We cannot offer advice here for lack of adequate space. The number of cells to be counted ($n$) depends on how often we are resigned to making a wrong decision. In biological studies a risk of 1 mistake in 20 inferences is generally accepted; this is the '0·05 (or 5 per cent) level of significance'. The count will usually be made on a dichotomous population, i.e. a population divided into cells which are in mitosis and those which are not. In the next chapter we shall have a similar situation with respect to cells which are radioactively labelled and those which are not. The appropriate statistical distribution for the treatment of observations drawn from such populations is the Binomial, but if $nI_M > 5$ we may use the Normal approximation. We shall *assume* kinetic uniformity (see Section 3.2.2).

Suppose we wish to have an estimate of the mitotic index which is sensitive enough to distinguish between $I_{M,1} = 0·01$ and $I_{M,2} = 0·005$ (say). We can determine the necessary size of the count $n$ by the following *approximate rule of thumb*. Let the variance of $I_{M,1} - I_{M,2}$ be

$$[(\hat{p}\hat{q}/n)_1 + (\hat{p}\hat{q}/n)_2]^{\frac{1}{2}}$$

with $p_1 = 0.01$, $p_2 = 0.005$. Now, assuming $n_1 = n_2$, we have

$$[(\hat{p}\hat{q}/n)_1 + (\hat{p}\hat{q}/n)_2] = 13.9 \times 10^{-3}/n$$

so that the standard error is $s_M = 0.1182/\sqrt{n}$. We now state an inequality, based on the Normal distribution, which $n$ (our cell count) must satisfy:

$$(0.01 - 0.005)/(0.1182/\sqrt{n}) > 2$$

i.e. $n > 2235$.

If, therefore, we tally 2300 cells or more, noting the mitoses *en passant*, we should have a sufficiently sensitive count. Since we are approximating a discrete distribution by a continuous one a correction should, in theory, be made. We omit it since it is of the order of $1.0 \times 10^{-4}$ and is therefore negligible. A more powerful procedure for the design of samples is discussed by Snedecor and Cochran (1971) which we commend to the reader.

## 3.2  Analysis by arresting mitosis in metaphase

Before the advent of cytokinetic techniques there was a tendency to regard the frequency of mitosis as a guide to the rate at which a tissue was proliferating. This criterion, however, is inadequate, even when expressed as a counted proportion of all cells in the particular region under investigation. It is inadequate because it is ambiguous; the number of mitoses seen depends (a) on the rate at which cells enter mitosis, but also (b) on how long they remain there. Thus if the duration of mitosis is long many mitoses will be seen even in populations which proliferate relatively slowly.

### 3.2.1  *Metaphase arrest (stathmokinesis)*
Probably the first technique to provide clear, if limited, information about the cell cycle was the use of substances which arrested dividing cells in metaphase. These compounds are often called *stathmokinetic agents*. The first of these, colchicine, was obtained from the autumn crocus (*Colchicum autumnale*). It is of some historical interest that the stathmokinetic use of colchicine was anticipated by Pernice (1889) who clearly illustrated metaphases, typical of colchicine treatment, in the intestinal mucosa of the dog. The significance of this observation was unrecognized for many years: in fact colchicine was not exploited as a metaphase arrest agent until the 1930s.

The stathmokinetic agents in use today are colchicine, its derivative Colcemid, and the increasingly popular vinca alkaloids vinblastine and vincristine. When added *in vitro* to a culture of proliferating cells, or given *in vivo* to laboratory animals, these drugs cause cells to accumulate in the metaphase stage of mitosis (Fig. 3.1) by interfering with the formation of the metaphase spindle. Therefore, if samples of the cell population under study are fixed at intervals over a period of several (e.g. four) hours, and the progressively rising metaphase index is deter-

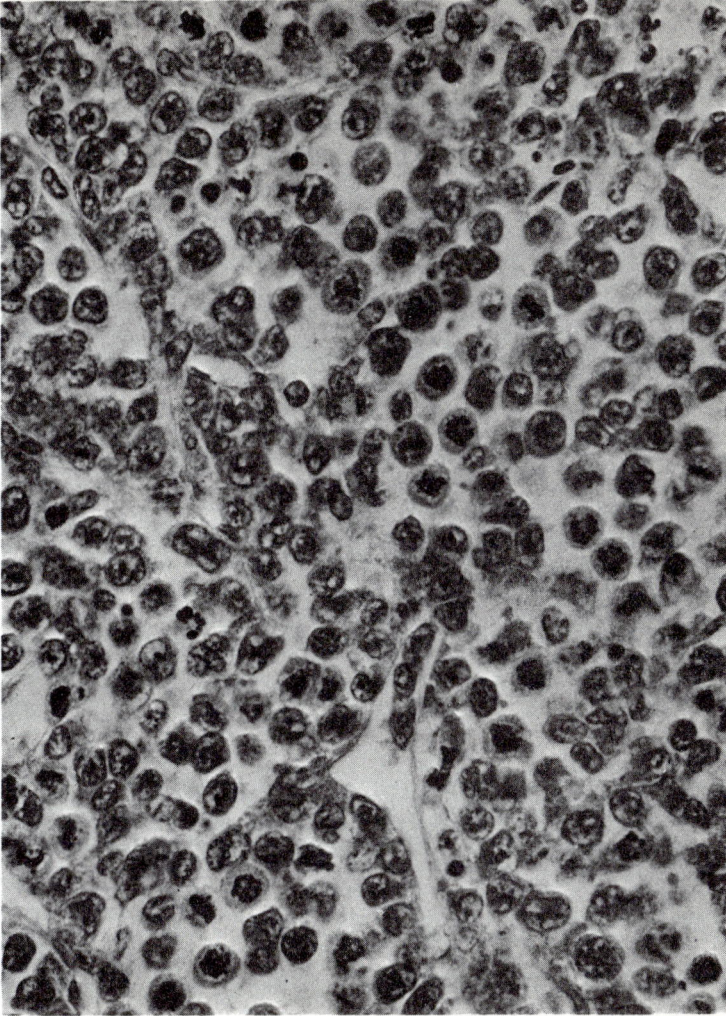

**Fig. 3.1** This photomicrograph shows the increased proportion of cells blocked in metaphase by the vinca alkaloid vincristine. The clumped state of the chromosome mass in most of the metaphases is characteristic. In a few cells the clump of chromosomes is degenerating: the loss of such cells leads to an underestimate of the arrested metaphase index after about 6 hours.

mined, then under ideal conditions the rate of entry of cells into mitosis, $r_M$, can be found. There are practical snags which may impede the estimation of $r_M$. These will be discussed shortly; for the moment let us *assume* that we can assess with sufficient accuracy the rate at which cells enter mitosis.

For tissue in an overall kinetic steady state the following relationship holds, on average:

$$r_M = I_M/t_M.$$

where $I_M$ is the mitotic index, and $t_M$ is the duration of mitosis. From this the stathmokinetic equation follows readily:

$$r_M = I_{met}/t_A \tag{3.1}$$

where $I_{met}$ is the metaphase index at the end of the arrest (or stathmokinetic) period $t_A$. In the case of exponentially growing populations, such as tumours, in which there is a bias towards young cells in the age distribution structure of the population, we use the equation derived by Puck and Steffen (1963):

$$r_M = \log_{10}(1 + I_{met})/t_A = 0.301/t_{C(a)}. \tag{3.2}$$

In Section 3.2.2 we shall consider the statistical aspect of $r_M$ in its role as a regression coefficient.

At mitosis one cell generates two daughter cells, therefore $r_M$ gives the birth rate of new cells, $k_B$. For steady state situations the experimental results are plotted in the form shown in Fig. 3.2(*a*). From this graph we can read not only $k_B$ but the average duration of mitosis, $t_M$, and the apparent cell cycle time, $t_{C(a)}$. The average duration of mitosis is

$$t_M = I_M/r_M \text{ (cf. Equation 4.1).} \tag{3.3}$$

The average apparent cell cycle time, $t_{C(a)}$, is given by the reciprocal of the slope of the line in Fig. 4.1(*a*). In the case of exponentially growing populations the experimental points are conveniently plotted on a semi-logarithmic grid, as shown in Fig. 3.2(*b*). The slope of the line is $r_M = 0.301/t_{C(a)}$, and its *position* may depend upon the length of time between the injection of the stathmokinetic agent and the beginning of metaphase accumulation. Thus both $t_M$ and $t_{C(a)}$ can be found. This description is based on the use of an *ideal* stathmokinetic agent, and it ignores statistical error.

Tannock (1967) defined the properties that an ideal stathmokinetic agent should have, namely: (a) at the dose level used and in the particular tissue under study, it should arrest *all* metaphases during the observation period; (b) these arrested metaphases should not degenerate into an unrecognizable state before the tissue is fixed and examined; (c) the metaphase arresting ability of the agent should not have adverse effects on interphase cells. If these criteria are satisfied the stathmokinetic agent will show a broadly peaked dose—response curve, from which an optimal dose can be chosen. This dose should give a well-defined, linear rate of accumulation of arrested metaphases over a reasonable period of observation. At a dose just below the optimum there may be a lag period before metaphases begin to accumulate. This need not alter significantly the slope of the line obtained in Fig. 3.2 but it will alter the intercept, and calculations of $t_M$ are particularly sensitive to such effects. The broken lines in Fig. 3.2(*a*) and (*b*) show what we expect when using doses of a stathmokinetic agent which allow cells already in mitosis to complete the process. When the agent has been added, cells in metaphase will begin to accumulate after a time interval roughly equal to the duration of mitosis. None of the stathmokinetic agents so far tested satisfy entirely the criteria set out above but vincristine has generally been found to come nearest to doing so.

Even when the best stathmokinetic agent is used a number of problems may arise. Care must be taken in the design of stathmokinetic experiments to avoid

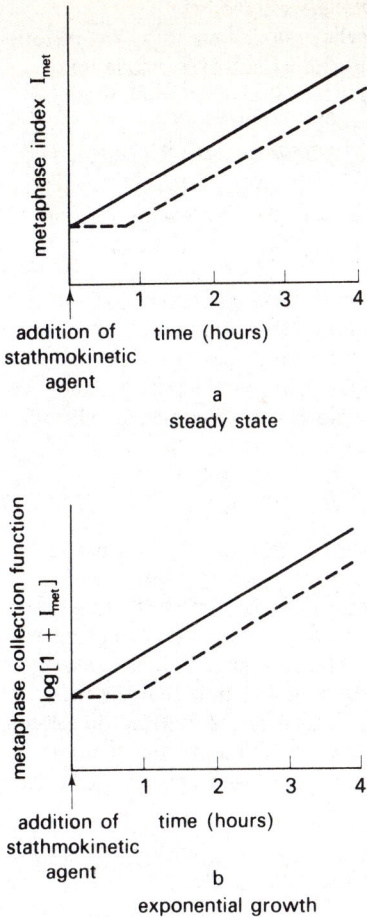

**Fig. 3.2** The collection functions derived by counting metaphases in successive samples from the population after an injection of a stathmokinetic agent: (*a*) in the case of the steady state; and (*b*) in the case of an exponentially expanding population. Note that (*b*) is linear on a semi-logarithmic plot. The significance of the broken lines is discussed in the text.

or correct for the degeneration of blocked metaphases, which are known to be functionally, morphologically, and biochemically abnormal and are therefore prone to break up and disappear (Aherne and Camplejohn, 1972). The life span of arrested metaphases tends to vary from tissue to tissue. Whatever stathmokinetic agent is used degeneration of blocked metaphases will occur if the period of observation is longer than five to six hours. Should this happen, or should an unsuitable dose of agent be used, then the mean birth rate ($k_B$) will be underestimated and this, in turn, will lead to an overestimation of the mean apparent cell cycle time ($t_{C(a)}$). By careful experimental design these untoward effects can be minimized.

### 3.2.2 *Some statistical aspects of mitotic and metaphase indices*

In Section 3.1 we saw that the population of cells in which we study the mitotic index is a dichotomous one, i.e. it is exclusively and exhaustively made up of cells in mitosis and cells in interphase. It follows that the mitotic index, like many other cytokinetic quantities, may be treated as a Binomial variate, *provided the tissue in question is uniform with respect to kinetic activity*. Unfortunately, this proviso is seldom, if ever, wholly satisfied. Let us consider (a) cases where a reasonable degree of kinetic uniformity prevails, and (b) cases where there are wide variations in kinetic activity from region to region. The latter is more re-alistic than the former (Tables 3.1 and 3.2).

In case (a) the block from which the histological sections are made represents the tissue in question as a whole. We therefore establish a confidence interval for the mitotic (or the metaphase) index, using Binomial statistics; this interval, we suppose, is true of the tissue as a whole. Since the count $n$ will be large, and $n\hat{p} \geqslant 5$ (where $\hat{p}$ estimates $I_M$), we may use the Normal approximation. Our 95 per cent confidence limits are therefore

$$I_M + 1 \cdot 96 \sqrt{(\hat{p}\hat{q}/n)} \text{ and } I_M - 1 \cdot 96 \sqrt{(\hat{p}\hat{q}/n)}, \tag{3.4}$$

where $\hat{q} = 1 - \hat{p}$. We interpret particular outcomes from a table of Normal values.

In case (b), where the mitotic activity varies from region to region, our histo-logical block cannot be assumed to represent the tissue as a whole. It represents *itself*; and therefore a confidence interval such as 3.4 applies to *counting* accuracy rather than to *sampling* accuracy. In case (b) we take as many blocks ($n$) as we can, subject to ethical and practical considerations, and assume that the mean values of $I_M$ are Normally distributed. Note that this is an assumption; in general we do not know the true distribution of $I_M$ values. We calculate the standard error of the mean mitotic index as usual. Then the upper and lower confidence limits are, conservatively,

$$\overline{I}_M + t(s/\sqrt{n}) \text{ and } \overline{I}_M - t(s/\sqrt{n}). \tag{3.5}$$

Similar situations, and some fresh problems, arise when we seek to establish a regression line through the scattergram of a stathmokinetic experiment. It will be instructive to consider again two contrasted cases. In case (a), the *homogenous* case, we suppose we have a tissue in which there is a uniform level of kinetic activity, from site to site in a given instance *and* from host to host. We have a choice of taking our serial samples from separate sites or from separate hosts. We might choose the latter if multiple sampling of the same specimen put its kinetic uniformity at risk.

Our variable is now $I_{met}$, i.e. the index of arrested metaphases. If, as usual, we base this index on a large count of cells (3000, say) we may assume the proper-·ties of the Normal distribution, as we did above. But the regression analysis may not be as robust as the *t*-test. Strictly, the experimental data may need to be transformed, since *proportions* do not satisfy the assumptions on which regression is ideally based. In particular, the mean and the variance are related. The appro-priate transform is the *arcsin-root*. This transformation can be carried out quickly by means of tables which are available in most statistical texts (e.g. Zar, 1974). It yields an approximately Normal distribution. In our context

**Table 3.1** The mitotic index, expressed as a percentage, at various sites in 5 cases of nephroblastoma

| Tumours | Sites | | | | | | | | | | | | Number of observations | Totals | Means |
|---|---|---|---|---|---|---|---|---|---|---|---|---|---|---|---|
| 1 | 1·08 | 0·32 | 1·04 | 0·97 | 0·60 | 0·44 | 1·43 | 0·96 | 1·26 | 0·39 | 1·06 | 1·06 | 12 | 10·59 | 0·883 |
| 2 | 0·63 | 0·38 | 0·60 | 0·64 | 0·44 | 0·75 | 0·75 | 0·43 | 0·66 | 0·69 | 0·70 | 0·67 | 12 | 7·30 | 0·609 |
| 3 | 1·42 | 0·75 | 1·04 | 1·45 | 0·94 | 0·16 | 1·23 | 1·00 | 1·09 | 1·10 | 0·80 | | 11 | 10·97 | 0·998 |
| 4 | 0·42 | 0·49 | 0·49 | 0·32 | 0·84 | 0·78 | 0·82 | 0·76 | 0·85 | 0·85 | 0·96 | | 10 | 6·61 | 0·661 |
| 5 | 1·16 | 1·08 | 1·00 | 0·84 | 0·64 | 1·33 | 2·11 | 0·83 | 1·10 | 0·70 | | 1·02 | 12 | 12·77 | 1·064 |

**Table 3.2** Analysis of the data shown in Table 3.1 — mean (%), variance, standard error, standard deviation, and coefficient of variation.

| Tumour | Mean (%) | $s^2$ | SE | $s$ | CV |
|---|---|---|---|---|---|
| 1 | 0·883 | 0·129 | 0·104 | 0·359 | 40·6 |
| 2 | 0·609 | 0·016 | 0·036 | 0·125 | 20·6 |
| 3 | 0·998 | 0·126 | 0·107 | 0·355 | 35·6 |
| 4 | 0·661 | 0·043 | 0·066 | 0·207 | 31·3 |
| 5 | 1·064 | 0·147 | 0·111 | 0·383 | 36·0 |

$$I_{met} \rightarrow I'_{met} = \arcsin \sqrt{(I_{met})} \sim N$$

where $I_{met}$ is the index of accumulated metaphases in a stathmokinetic experiment, and $I'_{met}$ is its Normally distributed transform, as the symbols $\sim N$ imply. It is an inherent weakness of this transformation that its efficiency falls off where it is most needed; in our case where the proportions are very small. It may also affect the linearity of the regression line.

The arcsin-transformed values of the metaphase index must, of course, be transformed back into proportions when we have finished our manipulations. The reader should be prepared to find that confidence limits may not be symmetrical about the mean when arcsin- ($\sqrt{}$) values are transformed back into proportions. Again, this reverse transformation is done by means of tables. But while we have our data in Normal form we may consider two further questions: how shall we decide whether an observed difference between two $r_M$ estimates is real or merely due to chance, and how much of the change in $I_{met}$ which we observe during a stathmokinetic experiment is due to a genuine accumulation of metaphases, as distinct from random variation?

The significance of the difference between two $r_M$ values is examined by calculating the *t*-value:

$$t = [\hat{r}_{M,1} - \hat{r}_{M,2}]/s_{rp} \tag{3.6}$$

where $s_{rp}$ is calculated from the pooled variances of $r_{M,1}$ and $r_{M,2}$.

If the value of $t$ calculated from **3.6** exceeds the tabulated value of $t$ for our chosen level of significance and $n_1 + n_2 - 4$ degrees of freedom then we accept the hypothesis that the two rates really differ. The second question may be answered by considering, in a similar way, whether the observed $r_M$ slope is significantly different from zero.

Now consider case (b), the *heterogenous* case. Here we suppose, as Tables 3.1 and 3.2 show for one particular tumour, that there is wide variation in kinetic activity from region to region in the same specimen of tissue and from host to host with respect to that tissue. We suppose also that it may be difficult to obtain more than a few samples from any one specimen of the tissue in question, for practical or (in human studies) ethical reasons. We are obliged, therefore, to pool results obtained on different hosts. In effect, we carry out the same stathmokinetic experiment *on M hosts*, from each of which we take *one sample* at time $t_i$ ($i = 1, 2, \ldots, M$), and then assume that the resulting estimates are equivalent to those we might derive from *M samples* of tissue in *one host*. But in practice we do not know how much kinetic variation exists in any one specimen of the tissue in question. Unpublished observations by the authors on certain neoplasms of man (mainly nephroblastoma and colorectal carcinoma) suggest that between-tumour variation and within-tumour variation are of the same order of magnitude, in these tissues at least. Thus we add between-tissue error to within-tissue error when we pool our estimates.

The binomial distribution of the metaphase index becomes irrelevant in circumstances like these. This is clearly seen in a symbolic argument. If we let $Y_i$ stand for the metaphase index, or some function of it such as $\log(1 + I_{met})$, in a given tissue at time $t_i$ in a stathmokinetic experiment we may write out the model thus

$$Y_i = a + bt_i + e_i + c_i; i = 1, 2, \ldots, M \tag{3.7}$$

where (1) $a$ and $b$ are estimates of the usual regression parameters;
(2) $c$ is the between-tissue variation (alternatively, between-hosts);
and (3) $e$ is the counting error, i.e. the error of a proportion.

We may not know what the distribution of $c$ is but we may reasonably suppose that the set of mean values from $M$ hosts is at least approximately Normal. Now, we can make the Binomial error $e$ as small as we please by counting sufficiently many cells. But $c$ is beyond our control and we know it to be relatively large. Thus $c$ overwhelms $e$ and the model becomes

$$Y_i = a + bt_i + c_i; c \sim N(0, \sigma_c^2); i = 1, 2, \ldots, M \tag{3.8}$$

which is the usual regression model, if $c \sim N(0, \sigma_c^2)$ is true.

It is possible to object to our model on the ground that the value of $Y_i$ must lie between 0 and 1, since it is a proportion, while $(a + bt_i + c_i)$ may theoretically take any value between $-\infty$ and $+\infty$ because of the tails in the distribution of $c$. This objection may be met by working with a transformation of $Y_i$ which does take values between $-\infty$ and $+\infty$. Two such transformations are the 'probit' and the 'logit'.

We have now arrived in difficult, poorly charted territory, where more research is needed. The reader who wishes to pursue the topics of the last few paragraphs in more detail should do so under the guidance of a statistician.

## Summary

To establish a reliable value for the mitotic index, or the metaphase index in a stathmokinetic experiment, it may be necessary to count a total of 2000 to 3000 cells.

The value of the mitotic index is only a rough guide to the proliferative activity of a tissue because it depends on the duration of mitosis as well as on the number (or proportion) of cells which are proliferating.

In a stathmokinetic experiment a specific drug (e.g. one of the vinca alkaloids) is used to stop mitosis in metaphase by wrecking the mitotic spindle. In this way we manufacture a kind of mitotic index whose duration is under the control of the experimenter; samples are taken from the tissue at precisely timed intervals $(t_A)$ after injection of the spindle poison. If the tissue is in a steady kinetic state the rate of entry into mitosis is given by $r_M = I_{met}/t_A$. If it is an exponentially expanding tissue the corresponding formula is $r_M = \log_{10}(1 + I_{met})/t_A = 0.301/t_{C(a)}$. Values for $r_M$, $k_B$, and $t_{C(a)}$ and $t_M$ can be read from a graph showing the accumulation of metaphases against time.

There are large stochastic errors in most stathmokinetic experiments, from two sources; a sampling error at any particular site, and a variation in proliferative activity from site to site in most tissues. Since we are dealing with a dichotomous population (cells in mitosis, cells not in mitosis) the appropriate statistical distribution is the Binomial; but if large numbers of cells are counted we can use the Normal distribution by approximation. A further source of error is the variation in the kinetic activity of a particular tissue (especially a tumour) between hosts. This error confronts us when we approximate a serial experiment on *one* host by a series of single experiments on *several* hosts.

## References

AHERNE, W. A. and CAMPLEJOHN, R. S. (1972) On correcting the error due to metaphase degeneration in stathmokinetic studies. *Exp. Cell Res.*, 74, 496.

PERNICE, B. (1889) Sulla caricincsi delle cellule epitheliali e dell'endotello dei vase della mucosa della stomach e dell'intestino, nella studio della gastro-enterite sperimentale (nell'arrelenamento pereolchico). *Sicilia Med.*, 1, 265.

PUCK, T. T. and STEFFEN, J. (1963) Life cycle analysis of mammalian cells. *Biophys. J.*, 3, 379.

SNEDECOR, G. W. and COCHRAN, W. G. (1971) *Statistical Methods*, 6th Edition. Iowa State University Press, Ames, Iowa.

TANNOCK, I. F. (1967) A comparison of the relative efficiencies of various metaphase arrest agents. *Exp. Cell Res.*, 47, 345.

ZAR, J. H. (1974) *Biostatistical Analysis*. Prentice-Hall, Inc., New Jersey, USA and Hemel Hempstead, UK.

# 4 How to analyse the cell cycle: II

**Techniques based on the use of radioisotopes**

Once the chemical composition of DNA and RNA was established it became clear that only three of the four component bases were common to both nucleic acids. Each nucleic acid contains adenine, guanine, and cytosine, but only DNA contains thymine. Due to the specificity of the nucleoside thymidine, in particular, a great stimulus to kinetic investigation came with the successful synthesis of tritiated thymidine ($^3$H-TdR) in 1957. A few years later $^{14}$C-thymidine ($^{14}$C-TdR) became available. Both tritium ($^3$H) and carbon-14 ($^{14}$C) are isotopes which emit $\beta$-particles.

## 4.1 Autoradiography

When either of these compounds is made available to proliferating cells those which are in the S-phase incorporate it and thus become recognizable as the S-cohort from then on. Labelled cells may be recognized by autoradiography (ARG). This entails placing a thin layer of photographic emulsion on sections of the tissue to be studied and leaving the sections in the dark for a period of days or weeks. When the photographic emulsion is developed those cells which contain radioactively labelled DNA in their nuclei will have black silver grains deposited in the emulsion above them (Figs. 4.1, and 4.2). These grains are caused by the radioactive emissions. Tritiated thymidine tends to be used for high resolution autoradiography in preference to $^{14}$C-thymidine because the silver grains are clustered more neatly over the labelled nuclei. This is so because $\beta$-particles emitted by tritium have less energy and therefore a shorter range than $\beta$-particles emitted by $^{14}$C.

## 4.2 The labelling index

Exposure of a cell population to labelled thymidine for a short period will enable the fraction of cells synthesizing DNA to be determined. This fraction is called the *pulse labelling index* ($I_s$). Its use as a guide to the proliferative activity of a cell population is analogous to that of the mitotic index and it is beset by similar problems. However, it does have a number of advantages: the labelling index is a larger fraction than the mitotic index, because the duration of the S phase is longer than that of mitosis, usually by a factor of 10 to 20. It follows that $I_s$ can

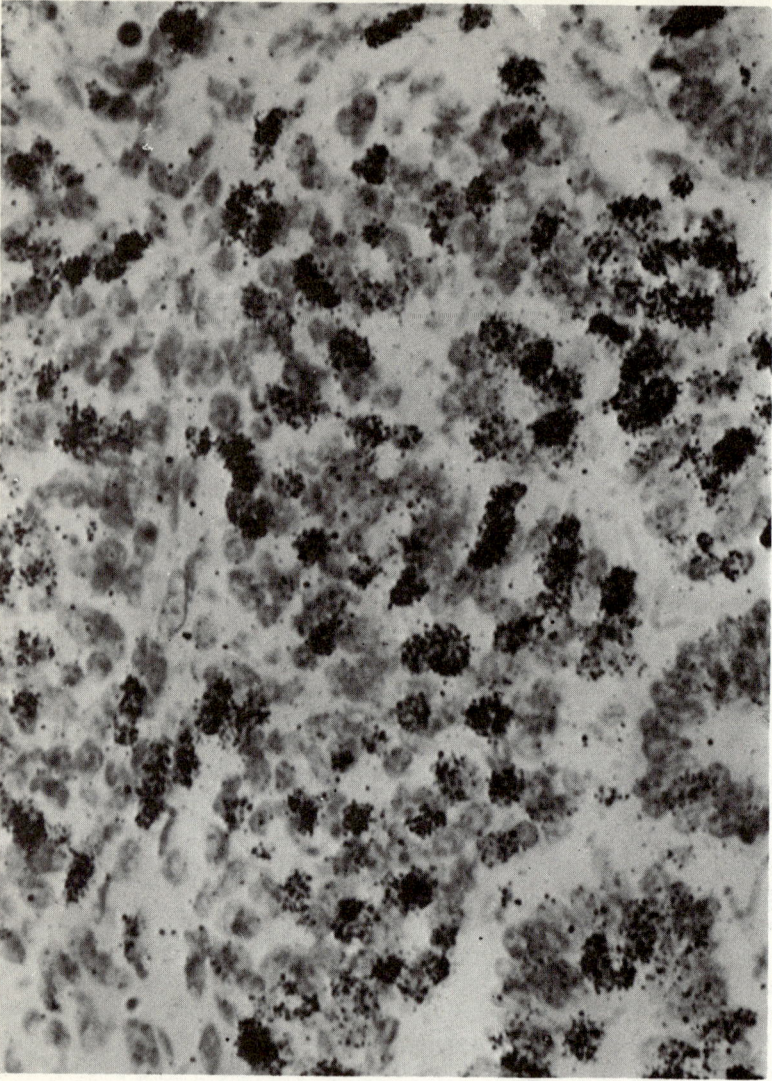

**Fig. 4.1** This shows an autoradiograph (at high magnification) of a rapidly proliferating embryonic tissue. Note the deposition of silver 'grains' in the photographic emulsion over nuclei which have incorporated $^3$H-TdR and are therefore emitting $\beta$-particles.

have a smaller relative error. Alone, the labelling index, though better than the mitotic index, still gives us only a rough guide to the proliferative activity of cell populations.

## 4.3 Double labelling techniques ($^3$H and $^{14}$C)

The double labelling technique enables the duration of the S-phase and the flux

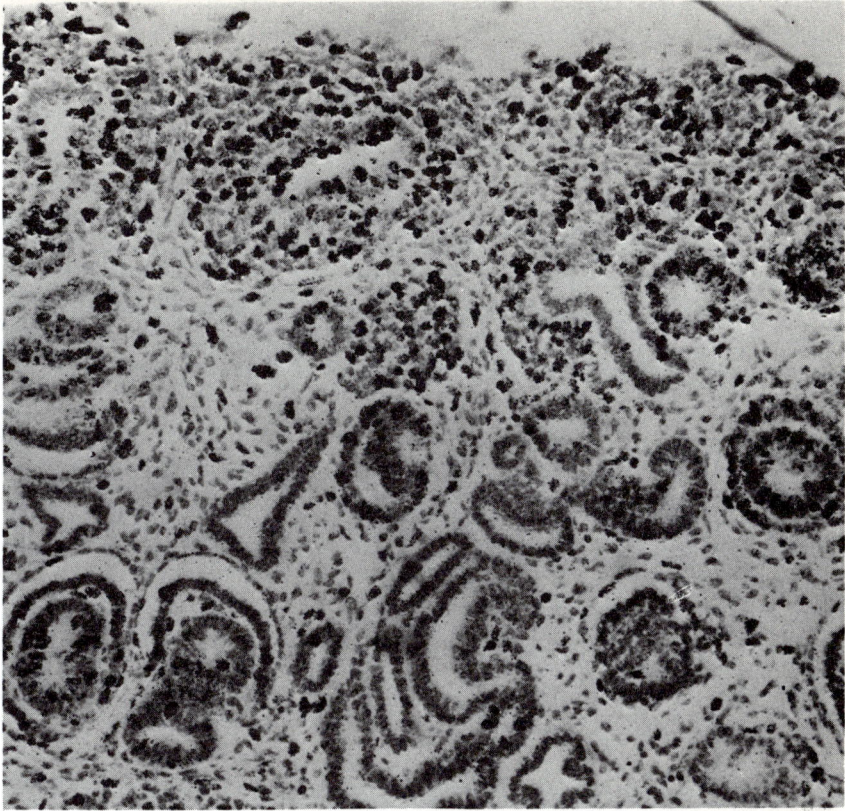

**Fig. 4.2** This shows an autoradiograph prepared by culturing a tissue (human fetal kidney) in a medium containing tritium-labelled thymidine, which is taken up and incorporated into DNA during the S-phase. The labelling is more intense towards the periphery of the tissue partly because this is the most actively growing region but also (as an artefact) because the available concentration of label and essential metabolites falls off as the diffusion distance from the medium increases.

rates into and out of S to be determined. This technique exploits the energy difference between the $\beta$-particles emitted by $^{14}C$ and those emitted by $^{3}H$. The procedure, in outline, is given below.

The cell population is first labelled with a pulse of $^{3}H$-thymidine; then, after an interval of about one hour, it is given a pulse of $^{14}C$-thymidine. Due to the higher energy and longer range of the $\beta$-particles emitted by $^{14}C$, comparatively widely-scattered *tracks* are seen in the photographic emulsion above $^{14}C$-labelled nuclei. Cells which have taken up $^{3}H$-thymidine *only* can be distinguished because they have a more localized deposit of silver *grains* above their nuclei. Generally, it is easy to distinguish cells which are labelled with $^{3}H$ alone from cells which are labelled with $^{14}C$, or $^{14}C + ^{3}H$, but it is difficult to distinguish nuclei labelled with $^{14}C$ only from those which have incorporated both $^{14}C$ and $^{3}H$. However, if two layers of photographic emulsion are used, separated by an *inert layer* of gelatin, or agar, the technique is greatly refined. Only $^{14}C$ emissions have enough energy and effective range to pass through the inert middle layer and precipitate silver tracks in the upper emulsion. Thus, by focusing the microscope first on the lower emulsion layer and then on the upper emulsion layer it becomes possible to distinguish three populations, namely: (1) cells which have incorporated $^{3}H$ only, (2) cells which have been labelled with both $^{3}H$ and $^{14}C$, and (3) cells which have incorporated $^{14}C$ only. If careful attention is given to technical details in general, and to emulsion thickness and the $^{3}H:^{14}C$ activity ratio in particular, all three populations can be easily distinguished.

By this double emulsion technique it is possible to measure both the rate at which cells enter the S phase (from the number of $^{14}C$-only labelled cells) and the rate at which cells leave the S phase (from the number of $^{3}H$-only labelled cells). This is illustrated in Fig. 4.3. In a steady state situation these two rates are

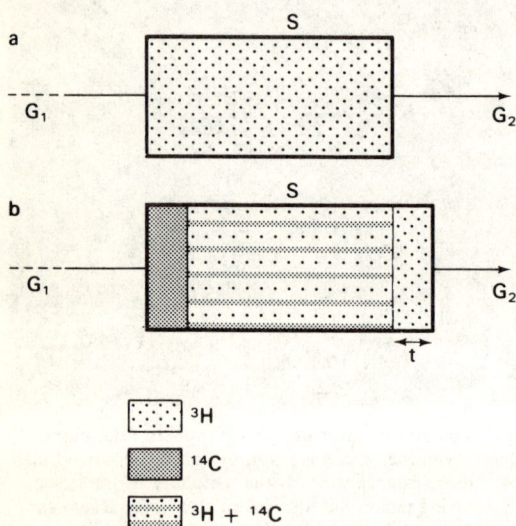

Fig. 4.3 A diagrammatic view of the double labelling technique. The rectangles represent the S-phase. The fine stippling shows the proportion of cells labelled by $^{3}H$-thymidine only, and $t$ is the time interval between the injections of $^{3}H$-thymidine and $^{14}C$-thymidine.

equal, but in an exponentially expanding system the rate of entry of cells is higher than the rate of exit; the difference can be as high as $r_{S(entry)}:r_{S(exit)} = 1 \cdot 5 : 1 \cdot 0$. In a steady state situation we can calculate $t_S$ from the equation

$$t_S = I_S/r_S \text{ (cf. Equation 4.7).} \tag{4.1}$$

The use of this equation in a non-steady state system will result in an error, the size of which depends on the kinetic characteristics of the system under study. In cell systems where the cell cycle time is fairly long (say 20 hours or above), the error is likely to be small. However, equations are also available for use in rapidly dividing exponential populations.

Instead of using $^{14}C$ and $^{3}H$ to distinguish cells labelled at different times, it is possible to achieve the same effect by giving two pulses of $^3$H-thymidine of widely different concentrations, again separated by an interval of time. Lightly labelled cells (from the first pulse) are distinguished from heavily labelled cells (from the second pulse) by the number of silver grains over their nuclei. A grain count distribution must be obtained for both lightly and heavily labelled cells (see Section 4.8), and the two doses of $^3$H-thymidine adjusted so that a good distinction between lightly and heavily labelled cells is obtained. Such a clear distinction is not always easily achieved, particularly with tumour cell populations. The use of the two isotope technique is preferable.

## 4.4  Statistical evaluation of labelling indices

Pulse labelling creates a population which is divided into two mutually exclusive and exhaustive classes. This is true of both single and double labelling methods. The statistical evaluation of the labelling index is subject to the same qualifications as the mitotic index, which we discussed in Section 3.2.2 (p. 21). In theory, the appropriate statistical distribution is the Binomial. But the variation in labelling, as in mitosis, usually renders Binomial statistics irrelevant, except as a measure of the error in a particular count on a particular histological section (cf. Equation 3.4). If a representative set of sections is available, and the count is large, the labelling index may be treated as a Normal variable. The 95 per cent confidence limits will then be

$$I_S + t(s/\sqrt{n}) \text{ and } I_S - t(s/\sqrt{n}) \text{ (cf. 3.5).} \qquad (4.2)$$

The significance of a difference between two mean labelling indices is tested by

$$t = (I_S^{(1)} - I_S^{(2)})/(s_p/\sqrt{n}) \qquad (4.3)$$

where $s_p$ is calculated from the pooled variances of $I_S^{(1)}$ and $I_S^{(2)}$ and one then refers to a table of Standard Normal Distribution values for the probability of the particular outcome in question.

## 4.5  The fraction of labelled mitoses (FLM) technique

This is one of the earliest and still one of the most informative of the isotope methods. It entails making a pulse of $^3$H-thymidine available to the cell population for a brief period of time: hence the term 'pulse', or alternatively 'flash' labelling. If the experiment is carried out *in vivo* the thymidine is rapidly removed by metabolism. If, on the other hand, it is done *in vitro* the $^3$H-Tdr must be removed by washing and any remaining $^3$H-Tdr diluted by the addition of 'cold' (i.e. unlabelled) thymidine. Tissue or cell samples are then removed and fixed — in a typical experiment — at hourly or two-hourly intervals for a period of about 50 hours after pulse labelling. Histological sections are prepared from these samples and the cells which have incorporated the labelled DNA precursor can be distinguished by the use of autoradiography.

Figure 4.4 shows the progress of the experiment. The cohort of $^3$H-labelled cells can be followed in the serial samples by noting the rise and fall of the fraction of labelled mitoses with the passage of time. Mitosis, unlike other phases of the cycle, is visually distinctive — it provides a window through which the pro-

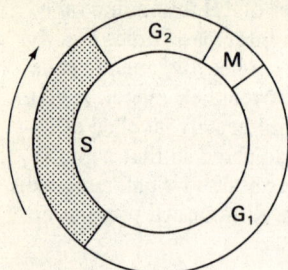

**Fig. 4.4** A diagram of the cell cycle partitioned into its phases. The stippled area in S represents a cohort of cells which have incorporated ³H-thymidine into their DNA. This is the situation when a population of proliferating cells is given a 'pulse' (or 'flash') label, i.e. when ³H-thymidine is made available for a brief period.

gress of the labelled cohort can be observed. Were there no variation in $t_C$, an identical peak would be generated each time the labelled cohort passed around the cycle and through mitosis. Such an idealized FLM curve is shown in Fig. 4.5. Figure 4.6 explains how this curve is generated by the passage of labelled cells around the cycle. This diagram also illustrates how we can obtain estimates of the duration of the cell cycle and its component phases.

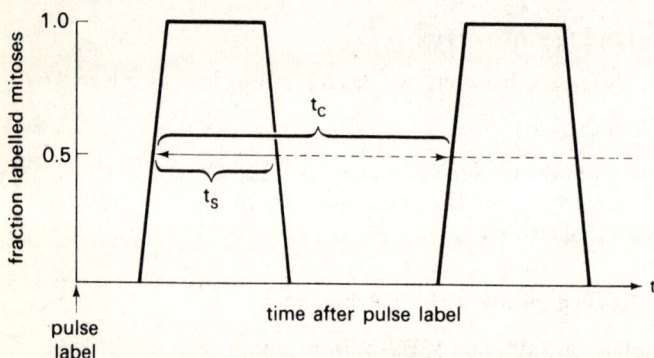

**Fig. 4.5** This idealized pair of FLM waves shows the time values of $t_S$ and $t_C$ by inspection. A graph such as this would only be possible if the system were deterministic, i.e. without the variation characteristic of biological systems.

Of course, in real proliferating populations the duration of each phase varies around a mean value. This causes damping of the FLM curve. The longer the time that has elapsed after pulse labelling, the greater will the effect of this damping be. A real FLM curve is illustrated in Fig. 4.7. It can be seen that the peaks are rounded and the second peak is more rounded than the first. Figure 4.8 illustrates a case where variation in cycle parameters, or the length of the cycle, or both, were such that by the time a second peak should have occurred the labelled cells had become completely dispersed around the cell cycle. Therefore no second peak could be obtained. The durations of the cell cycle and its phases are usually

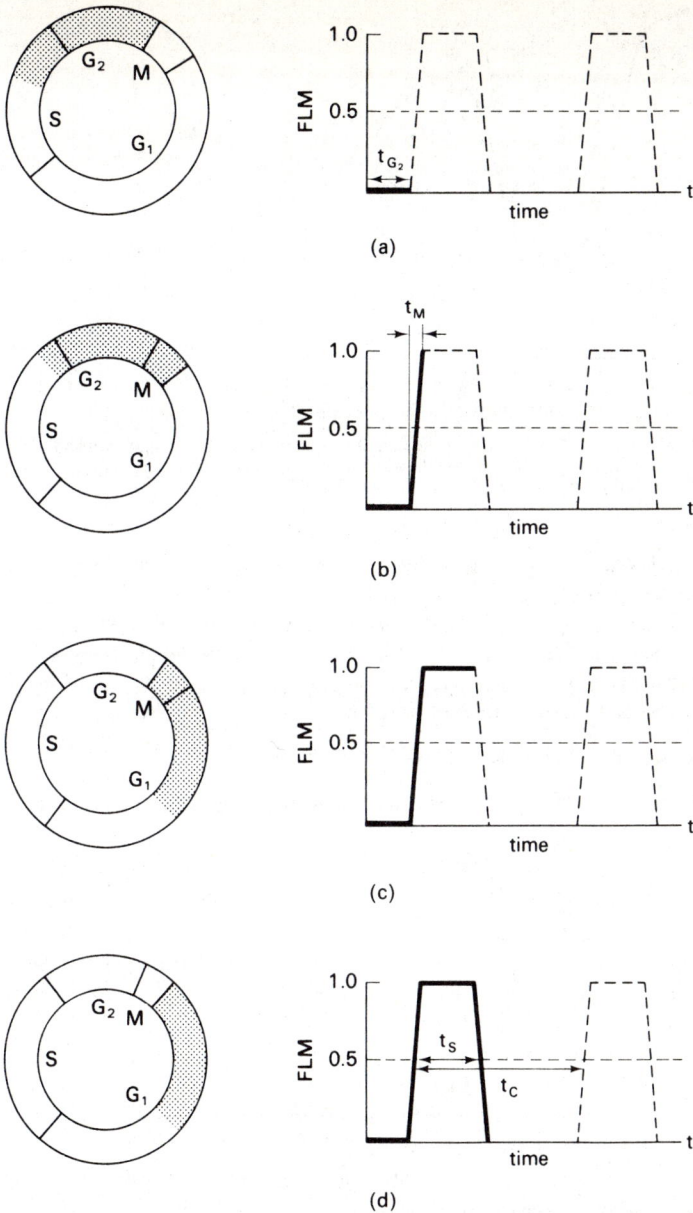

**Fig. 4.6** On the left of the diagram a cohort of labelled cells is depicted, moving through the cell cycle. As it moves into M one sees an increasing proportion of labelled mitoses, as shown on the right of the diagram. The time interval from the injection of $^3$H-thymidine to the median of the ascending limb of the idealized first wave measures $t_{G_2} + (1/2)t_M$. When the cohort of labelled cells has occupied M completely, as at (*b*), the proportion of labelled mitoses is unity (or, as it is sometimes stated, 100 per cent). This is represented on the right by the peak of the idealized first wave. In (*d*) the cohort is moving away from M into $G_1$. When it reappears the duration of the cycle ($t_C$) will be measured by the interval from the median of the first wave to the median of the second.

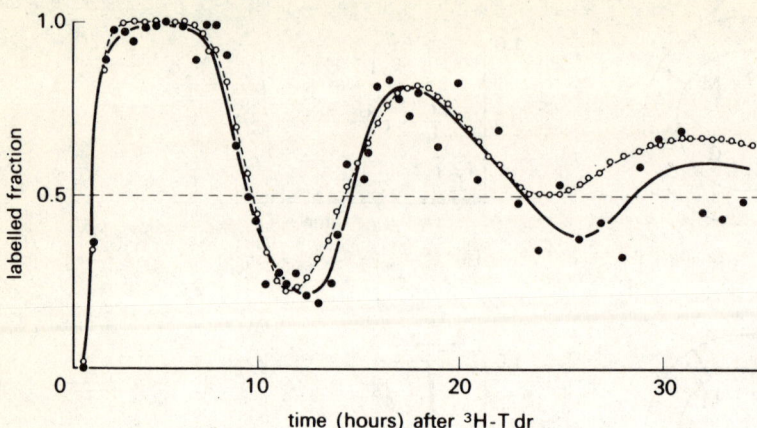

Fig. 4.7 A real FLM curve, derived from studies on the murine small intestinal mucosa. Though there is clear evidence of damping it is still possible to derive $t_S$ and $t_C$ at least. The dotted line was generated by a computer using Gilbert's (1972) programme.

read from the 0·5 levels. This gives values approximating to the median values for the population under study.

Barlow and Macdonald (1973) have published a method of analysing incomplete FLM curves such as that depicted in Fig. 4.8. Its validity depends upon good estimates of the mitotic index $I_M$, and the labelling index $I_S$. We state their essential formula with a slight change of notation:

$$t_C = [-t_S \ln p] / [\ln\{1 - (p-1)I_S p^{-0.5a-b/t_C}\}] \qquad (4.4)$$

where $t_S$ (hr) is the width of the FLM wave, measured at the 0·5 level, $p$ is the

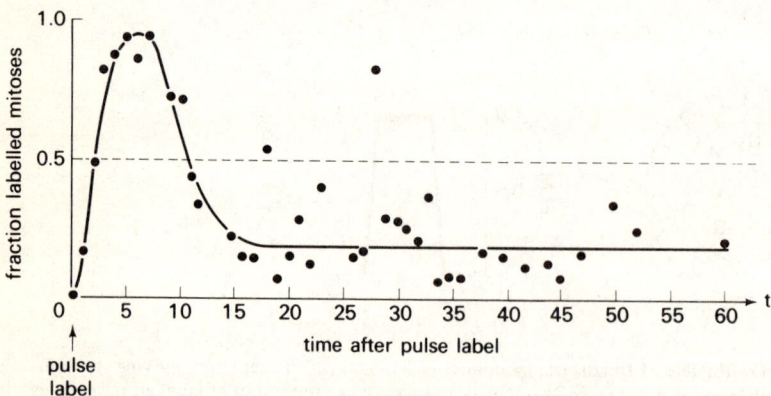

Fig. 4.8 Another real FLM curve, derived from studies on a pulmonary adenoma induced in the mouse by urethane. Damping is severe; in fact there is no clear second wave. In these circumstances the eventual plateau level may be equated to $t_S/t_C$, and thus $t_C$ may be found if there is a first wave from which a satisfactory estimate of $t_S$ can be measured (by courtesy of Dr. Paul Dyson and Professor A. G. Heppleston).

mean number of daughter cells which continue to proliferate after a mitosis (i.e. $p \leqslant 2$), and

$$t = t_{G2} + 0.5\, t_M$$

$$a = \ln\{1 + (p - 1)I_M\}\,(\ln p)^{-1}$$

$$b = t + t_S - \tau.$$

Here $\tau$ is the pulse length, i.e. the period of time for which the label ($^3$H-TdR) is available to the cells. If we assume that $p = 2$ and $\tau \approx t$ we have a simpler and more familiar form:

$$t_C = [-t_S \ln 2]\,/\,[\ln\{1 - I_S 2^{-0.5a - b/t_C}\}] \tag{4.5}$$

$$a = \ln(1 + I_M)/\ln 2$$

$$b = t_{G2} + t_S.$$

Equation 4.4 (or 4.5) has $t_C$ on both sides and so must be solved recursively by using trial values of $t_C$ until equality is established. If the FLM curve settles on a plateau level, $L$, a starting estimate of $t_C$ may be found from

$$t_C = t_S/L. \tag{4.6}$$

A number of workers have replaced the visual analysis of FLM curves by computer simulations of the cell cycle to generate a best fit curve for a particular set of FLM values. By assuming certain types of statistical distributions for the range of cell cycle parameters within a given population (e.g. normal, log-normal, etc.), the durations of the cell cycle phases and their variances can be calculated. Ideally, such statistics should tell us much more about the cell population in question, and enable us to compare FLM curves for significant differences. But simulation is a fairly recent development and the full validity of standard errors (for example) has not yet been rigorously established.

## 4.6 Continuous labelling

The next kinetic technique to be described also exploits $^3$H-thymidine but in a rather different way. Instead of making the label available for only a short period the cell population under study is subjected to labelled thymidine continuously for a long period. This can be achieved *in vivo* (in experimental animals, of course) either by repeated injection at periods shorter than the S phase, so that no cells going through S escape labelling, or by continuous infusion. These procedures may, however, subject the animals to a degree of stress undesirable for both physiological and humanitarian reasons. The technique of continuous labelling is probably best suited to use *in vitro* where the label can be added without perturbing the culture. Samples of the cell population under study are removed and fixed at intervals which will depend upon the characteristics of the particular cell population and upon the aims of the particular experiment. The labelling index ($I_L$) is determined, at each time point after addition of $^3$H-TdR, by counting a suitable number of cells and recording the proportion that have taken up the label. A 'suitable number' depends on the size of the standard error which the experimenter will accept.

At the beginning of the experiment those cells already in the S phase will take up the tritiated label. So will all cells entering S from then on, until a plateau level of labelled cells is reached. If all the cells in the population are in cycle (i.e. if the growth fraction is unity) then, as shown in Fig. 4.9, the entire population

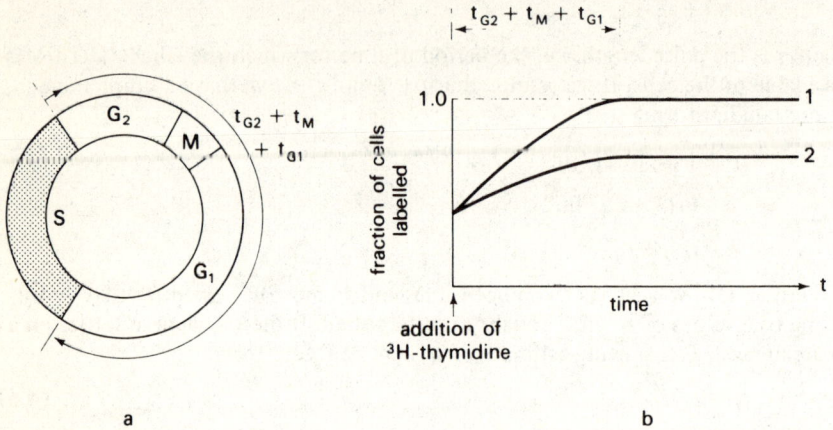

Fig. 4.9 Continuous (or repeated) labelling. If all of the cells in the population are proliferating, serial counts will give increasing labelling indices falling along line 1 in (*b*); if only a fraction of the population is in cycle then the serial labelling indices fall along line 2. It may then be difficult to decide exactly when a plateau has been reached. Conversely, and with less uncertainty, continuous labelling may be used to establish the growth fraction.

will be labelled in a time equal to the mean durations of $t_{G2} + t_M + t_{G1}$. Therefore, if we plot the fraction of labelled cells versus time after addition of label (Fig. 4.9*b*, line 1), an estimate of the combined durations of $t_{G2} + t_M + t_{G1}$ can be made. If, on the other hand, the growth fraction is significantly less than unity (line 2, Fig. 4.9*b*), this technique is blunted somewhat by the difficulty of deciding exactly when a plateau level has been reached. Conversely, and with less uncertainty, if observations can be sufficiently prolonged, continuous labelling can be used to *determine* the growth fraction.

In theory the time span $t_{G2} + t_M + t_{G1}$ can be broken down into its component intervals. However, in real cell populations which show variation in kinetic parameters and in which cell death occurs, the values obtained are only approximate and are difficult to enclose within statistical limits. The interval $t_{G2} + t_M$ is the time needed to label all of the mitoses (Fig. 4.10). This can be estimated by plotting the fraction of labelled mitoses in sequential samples. The value of $t_{G1}$ follows obviously from

$$(t_{G2} + t_M + t_{G1}) - (t_{G2} + t_M) = t_{G1}$$

The duration of the S phase ($t_S$) can be estimated in three ways from a continuous labelling experiment. The most straightforward of these is from the rate of increase in the fraction of labelled cells with time (i.e. the rate of entry of cells into DNA synthesis). This is analogous to the method described for calcula-

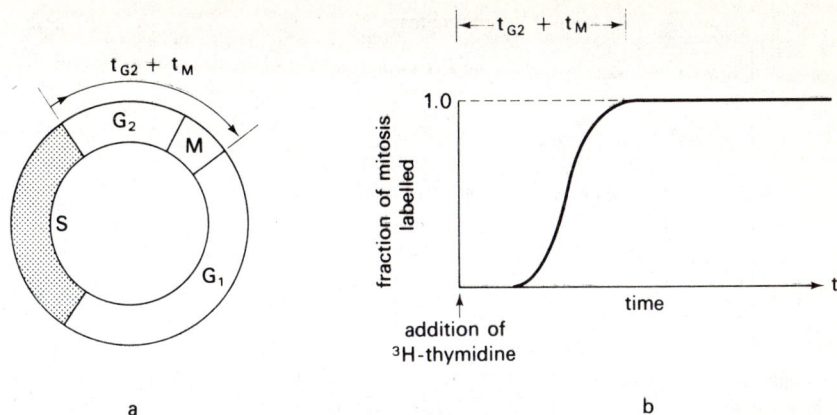

Fig. 4.10  The estimation of $t_{G_2} + t_M$ by continuous (or repeated) labelling. In this experiment the proportion of labelled mitoses is established in serial samples. The interval $t_{G_2} + t_M$ is estimated as the time needed to label all of the mitoses.

ting the duration of mitosis from stathmokinetic experiments and is calculated in an analogous way (see p. 20). It is also subject to the same limitations. The rate of entry into S is obtained from the first part of the graph in Fig. 4.9b, preferably before any labelled cells have had time to divide. The duration of S is then calculated from the formula

$$t_S = I_S/r_S \qquad\qquad (4.7)$$

where $r_S$ = the slope in the steady state case. In expanding populations $t_S$ is more difficult to calculate and it is best done using computer simulation techniques. It may be estimated roughly by the equation

$$t_S = I_S\, t_{C(a)}/\ln 2. \qquad\qquad (4.8)$$

Two other methods of estimating $t_S$ are possible using continuous labelling data; one involves a graphical technique using the age distribution diagram (Fig. 2.4); the other entails grain counting, a technique that will be described presently.

By adding the estimate of $t_S$ to that of $t_{G_2} + t_M + t_{G_1}$, an estimate of $t_C$ is obtained. This technique enables the growth fraction to be measured also (see Section 2.2, p. 9).

## 4.7  Combined metaphase blockade and continuous labelling

A very elegant technique due to Puck and Steffen (1963) combines stathmo-kinesis and continuous labelling. This technique is ideally suited to use in tissue culture where tritiated thymidine and a stathmokinetic agent can be added to culture media, and samples of tissue fixed over a period of hours. The number of mitoses, both labelled and unlabelled, and total labelled cells, are determined in each sample and the accumulation with time of all three components is then followed. For both steady state and exponential systems $k_B$, $t_M$, and $t_{C(a)}$ can be estimated, as we described in Chapter 3, from the rate of accumulation of mitoses.

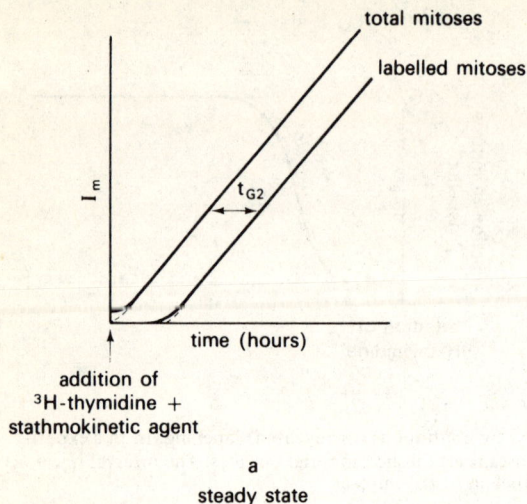

a
steady state

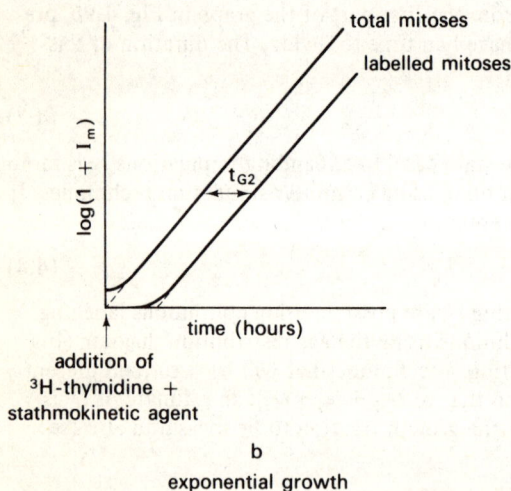

b
exponential growth

**Fig. 4.11** The technique devised by Puck and Steffen (1963) in which the accumulation of unlabelled mitoses, labelled mitoses, and total labelled cells is determined in serial samples. The linearizing function $\log(1 + I_M)$ is generally known as the 'collection function'; see text, Section 4.7. Note that allowance must be made for growth fractions less than unity.

If, in addition, the *collection function* of labelled mitoses is plotted (Fig. 4.11: (*a*), steady state; (*b*), exponential), $t_{G2}$ can be found. The duration of S (i.e. $t_S$) can be estimated from the accumulation of labelled cells as described in Section 4.6 for steady state conditions. For exponentially growing populations, the data may be plotted as shown in Fig. 4.12, in which case the intercept is equal to

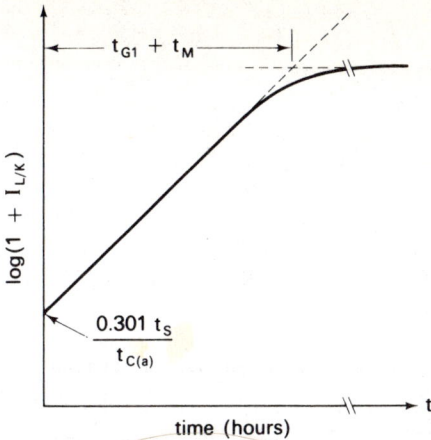

Fig. 4.12 The accumulation of labelled cells in an exponentially growing population in the presence of $^3$H-thymidine and a stathmokinetic agent. The collection function due to Puck and Steffen is plotted on the vertical ordinate. In the expression $\log(1 + I_L/K)$ $I_L$ is the fraction of the population which is labelled, and $K = \exp(\ln 2t_{G_2}/t_{C(a)})$. Note that allowance must be made for growth fractions less than unity.

$0·301$ $t_S/t_{C(a)}$; thus $t_S$ can be calculated. If the experiment is carried on until a plateau level is reached a direct estimate of $t_{G1}$ can be made (Fig. 4.12). Alternatively, if the growth fraction is known $t_{G1}$ can be calculated from:

$$t_{G1} = t_C - (t_M + t_{G_2} + t_S). \tag{4.9}$$

The mathematical basis of the equations given above is in a paper by Puck and Steffen (1963) and is also described by Cleaver (1967). It will not be discussed further here. In practical use in ideal rapidly growing systems this technique has been used successfully to resolve events in the cell cycle within fairly accurate limits. It can, however, be difficult to adjust the level of labelling so as to identify all labelled cells without obscuring mitoses.

## 4.8  Grain counting

So far, in the context of labelling experiments, the techniques described have involved determining the proportion of labelled interphase cells or labelled mitoses. In certain circumstances information can also be obtained by assessing the *intensity* of the labelling, and how that intensity varies with time during an experiment. Intensity of labelling can be determined autoradiographically by counting the number of silver grains over labelled nuclei. The mean (or median) number of grains per nucleus may be used to estimate certain parameters of the cell cycle. The usual procedure is to count the number of grains over a suitable number of nuclei in each sample examined and then to plot the distribution of counts as shown in Fig. 4.13. Two examples of the use of grain counting follow.

In experiments where a pulse label is given, such as the FLM technique, grain counts on the interphase (non-mitotic) cells are sometimes employed to estimate $t_C$. It is assumed that the number of grains over an individual cell is halved at

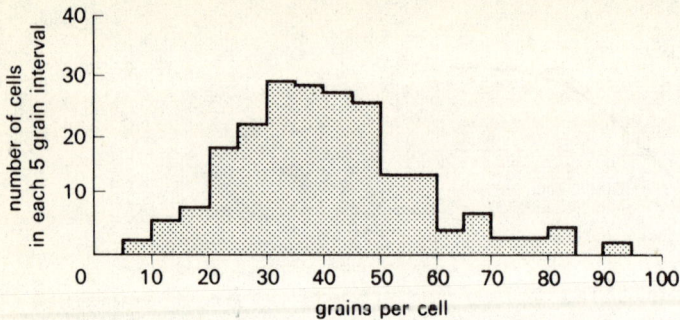

**Fig. 4.13** Histogram showing the frequency distribution of grain counts in a population of labelled cells. By deducing the interval between successive halving of the grain count an estimate of the cell cycle time ($t_C$) can be made.

mitosis. This being the case the halving time for the mean (or median) grain count should indicate the duration of the cell cycle. However, the first halving of the grain count occurs in a time which is less than $t_C$, because the cells have only to pass through S + G$_2$ + M for the first halving to occur. This of course will happen in a time equal to $t_S + t_{G2} + t_M$. After this, however, halving should take place at intervals of a cell cycle time. Thus, if mean grain count is plotted against time, $t_C$ can be estimated. The estimate obtained for $t_C$ is nothing like as accurate or reliable as that obtained from an FLM curve. Nevertheless the method can be used in cases where mitotic figures are rare and construction of an FLM curve is therefore difficult. Practical and technical difficulties may give rise to false grain counts, but a more fundamental problem arises when the technique is applied to tumours and other cell populations where there is a significant transition of cells from the proliferating to the non-proliferating state. These cells would cease to halve their grain counts and so lead to an overestimate of the halving time, and in turn, to an overestimate of $t_C$. This situation applies in most tumour cell populations. Published estimates of $t_C$ in tumours using this technique may well be too long.

Grain counting can also be of use in providing a method of estimating $t_S$ from a continuous labelling experiment. The maximum number of grains over a nucleus occurs when the whole genome has been replicated. This is achieved in a time interval equal to $t_S$ (Fig. 4.14).

Finally, the decycling probability (see Section 2.3.2) can be estimated by grain counting during an FLM experiment. As we saw above the mean grain count halves with each mitosis, provided all new cells re-enter the cycle. Decycling causes – and can be estimated from – a deviation from this halving sequence.

### 4.9 Scintillation counting

It is possible to assess the tissue uptake of compounds such as $^3$H-TdR and $^{14}$C-TdR by means other than autoradiography. Tissue samples are usually rendered soluble by a compound such as the hydroxide or chloride of Hyamine 10-X. By mixing the solubilized tissue sample with an organic liquid scintillator we can assess the amount of radioactivity in our sample. An organic scintillator consists

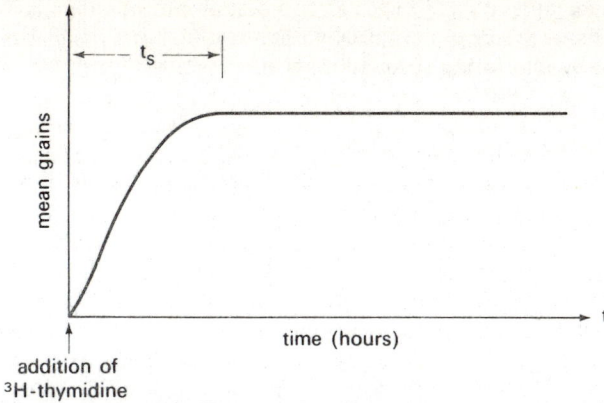

**Fig. 4.14** Grain count graph from a continuous labelling experiment. The count reaches a maximum when the whole genome has been replicated, and thus an estimate of the duration, $t_S$, of the DNA synthesis time can be made.

of one or more fluorescent aromatic solutes in an aromatic solvent. When the molecules of the fluorescent compound are struck by an ionizing particle (such as a tritium or $^{14}$C $\beta$-particle) they emit a light flash. We can count scintillations in our sample by means of a liquid scintillation counter. In this instrument the scintillations are detected by a photomultiplier tube, a sensitive device which converts weak light signals into amplified electrical pulses. The numbers of pulses per minute (i.e. the counting rate) is proportional to the number of original scintillations per minute, which in turn is determined by the number of disintegrations per minute (the 'activity') of the source of ionizing radiation.

The advantages of scintillation counting are these: (1) it is possible to obtain results much more quickly than is possible with autoradiography, since there is no step comparable in duration with the long exposure period for ARG; and (2) radioactivity can be measured in a large amount of tissue without the need for visual scanning and counting. Thus in some circumstances, for example in a culture of lymphocytes, where we wish to assess the proliferative response to a number of concentrations of an antigen, scintillation counting will give a much quicker result for much less work than counting labelled cells on autoradiographs can. However, if we have a tissue composed of different cell types, scintillation counting cannot, of course, tell us which cells are labelled. It may be very important to know which cells have bound the radioactive agent. Secondly, since scintillation counting measures radioactivity in a global and indiscriminate way changes in uptake may be due either to a change in the *number of cells* taking up labelled thymidine (in kinetics this is usually what we study), or to a change in the *amount of thymidine* taken up by each cell. The second possibility may occur due to changes in the sizes of thymidine metabolite pools within the cell. With autoradiography we can assess directly the number of cells taking up labelled thymidine.

## 4.10 Microdensitometry

This is a technique used to estimate the quantity of DNA in a given cell nucleus.

Combined with autoradiography it can provide a good deal of information relatively easily. The basic tissue preparation is usually not a section but a population of whole cells, flattened by imprinting or squashing tissue fragments gently on microscope slides. The fixed whole cells are then stained with a specific DNA stain such as Feulgen's, which is not only specific for DNA but is also quantitative; in other words the density of staining varies directly with the quantity of DNA present in the nucleus. By using an instrument that measures the amount of light absorbed by the Feulgen stained nuclei at a specific wavelength one can tell which cells contain 2n DNA ($G_1$ or $G_0$ cells) and which contain 4n DNA ($G_2$ cells). S-phase cells can be recognized and excluded by combining the technique with pulse labelling and ARG. In a population where all cells are proliferating the number of cells in all phases of the cycle can then be ascertained; S cells by pulse label, M cells visually, and $G_1$ and $G_2$ by densitometry. Thus the age distribution diagram (Chapter 2) for the population can be constructed. The relative durations of the phases of the cycle can be read from the abscissa of this diagram. If the duration of one of the phases, say $t_S$, has been measured independently then the absolute durations of the other phases can be calculated. But, as always, there are limitations. In tumours, and other cell populations where there are non-proliferating cells; $G_1$ and $G_0$ cells are indistinguishable by this technique, and the phase durations cannot be directly measured. Furthermore, again especially in tumours, cells which contain variable amounts of DNA (polypoid and aneuploid cells) cause difficulties. The technique nevertheless provides useful information in conjunction with other methods, in particular the FLM technique.

A relatively recent development is DNA microfluometry. In this technique a well separated cell suspension is stained with a dye which reacts stoichiometrically with DNA, such as propidium iodide. The cells are then passed through an instrument which, essentially, records the amount of DNA per nucleus in the form of a histogram. A distinct advantage of this 'flow through' technique is the large number of cells which can be quantified in a very short time. The proportion of the population in $G_1 S_1$ and $G_2/M$ can be calculated.

The main drawbacks of the technique are: (a) a good cell suspension can be difficult to obtain, for example in structured epithelial populations; (b) there is considerable debate over the mathematical analysis of the DNA histograms; and (c) results can be misleading in the case of tissues where proliferative rates vary with position in the microarchitecture. Nevertheless, the technique is certain to grow in popularity as the analytical methods improve.

## 4.11  Cell damage due to self-irradiation

[3]H-TdR can, if given in sufficient quantity, cause observable radiation damage in cells which have incorporated it. In pulse labelling studies it is generally assumed that a dose of 1 $\mu$Ci/g of body weight does not produce alterations in the cell kinetic parameters of cell populations. However, even this assumption may not always be justified. Certainly, higher doses may cause kinetic perturbations resulting in false estimates of normal kinetic parameters, and the danger is particularly acute in continuous labelling studies. This danger should be kept in mind and the minimum dose practical for the experimental purpose should always be used.

## Summary

There are many cytokinetic techniques based on the incorporation of tritium-labelled thymidine ($^3$H-TdR) into DNA during the S-phase of the cell cycle. For histological work sections are cut from fixed tissues which have incorporated $^3$H-TdR during life; a photographic film or liquid emulsion laid over the section detects the sources of radioactivity (i.e. the labelled cells) as clumps of photographic grains. This technique is known as autoradiography. The proportion of labelled cells in the population is the *labelling index*, $I_S$. Another isotope occasionally used, especially in combination with $^3$H, is radioactive carbon, $^{14}$C; together they provide an excellent method of estimating the duration of the S-phase. The labelling index has properties similar to those of the mitotic index; for example, $I_S = t_S r_S = t_S/t_C$ (cf. Equation 2.3): note, however, that $r_S$ in an exponentially expanding population is not equal to $r_M$.

The single most useful cytokinetic method is the 'fraction of labelled mitoses' or FLM procedure, in which, after a single injection of $^3$H-TdR (a 'flash' or 'pulse'), two or possibly more waves of labelled mitoses may be recorded as the labelled cohort of cells (i.e. those that were in S when $^3$H-TdR was given) traverse the cell cycle. The width of a wave of half-height estimates $t_S$, and the distance from the first wave to the second, again at half-height, estimates $t_C$. Computer programmes are available for the optimal plotting and statistical analysis of such data.

$^3$H-TdR may be supplied continuously (in the case of tissue culture) or by repeated injection (in the case of the experimental animal) to the cell population at issue. Continuous labelling methods may be used to establish $t_{G2} + t_M + t_{G1}$ and the growth fraction $I_P$ (or $f_G$).

Radioactive labelling may be combined with metaphase blockade to provide a particularly elegant method of deriving a number of cytokinetic parameters from one experiment.

Information about the rate of DNA synthesis can be derived from serial counts of the number of photographic grains over labelled cells. Counting grains also enables one to determine the duration of the cell cycle, since the average number of grains is halved when mitosis occurs.

When it is not necessary to establish which cells are labelled, i.e. when it suffices to know the total amount of radioactive precursor incorporated into a unit mass of tissue, the technique of scintillation counting is available.

## References

BARLOW, P. W. and MacDONALD, P. D. M. (1973) An analysis of the mitotic cell cycle in the root meristem of *Zea mays. Proc. Roy. Soc. (London)* B, **183**, 385.

CLEAVER, J. E. (1967) *Thymidine metabolism and cell kinetics*. North Holland, Amsterdam.

DYSON, P. and HEPPLESTON, A. G. (1976) Cell kinetics of urethane-induced murine pulmonary adenoma: II the growth fraction and cell loss factor. *Br. J. Cancer*, **33**, 105.

GILBERT, C. W. (1972) The labelled mitoses curve and the estimation of the parameters of the cell cycle. *Cell Tiss. Kinet.*, **5**, 53.

PUCK, T. T. and STEFFEN, J. (1963) Life cycle analysis of mammalian cells. *Biophys. J.*, **3**, 379.

# 5 Cell populations; theory and practice

So far we have considered some of the various parameters which are used to characterize populations. We must now move on to another plane and discuss populations as entities in themselves. In this chapter we will attempt to define the various types of real cell populations and, further, to break down and analyse these real systems in terms of their component subpopulations.

## 5.1 The relevance of population analysis

When studying any cell population, it is important to define as accurately as possible the population and its subdivisions in morphological terms. For example, proliferating cells may be located in a specific anatomical site, and by defining this a good idea of the spatial boundaries of the proliferative and non-proliferative subpopulations can be obtained. In squamous epithelium, labelled cells and mitoses are usually confined to the basal layers; in the crypts of Lieberkühn of the small bowel mucosa proliferative activity is found in the lower two-thirds of the crypt. Bone marrow presents a special problem; there are no distinct *spatial* boundaries between the proliferating and non-proliferating cells. Marrow subpopulations are defined here in terms of cell morphology.

It is also important to remember that cells often migrate from one anatomical site to another. Basal epidermal cells move upwards into the malpighian layer and crypt cells in the small bowel move upwards on to the villi. In this way cells move from the proliferative into the non-proliferative area. Bone marrow cells move gradually into the non-proliferative subpopulation as they differentiate towards mature red and white blood cells.

## 5.2 Classification of cell populations: I

Leblond and his colleagues (1959) proposed a simple classification of rodent tissues based on the extent to which they take up and retain injected thymidine labelled with tritium ($^3$H) as a DNA marker. This 'uptake and retention' of $^3$H-TdR was studied by preparing autoradiographs in rats of different ages within a few hours of injection; young rats were also labelled and killed 6 months later. Thus it was possible to study both cell formation and local retention. The classification arising out of these investigations may be summarized as follows:

(a) *stable cell populations*, which take up and retain the label early in their lifespan but show no cell formation in adults. Proliferation occurs early in life and the cells produced remain within the population. This group includes smooth and striated muscle and many types of neurone;

(b) *renewal cell populations*, which proliferate continuously and lose cells continuously in a dynamic equilibrium, or steady state. This group includes bone marrow, small bowel mucosa, and epidermis, when these tissues are considered over all.

(c) *expanding cell populations*, which are capable of both cell proliferation and cell retention through adult life. Liver parenchymal cells and renal tubular cells are examples of this category. Excision of part of the liver is followed by a resurgence of mitotic activity in the remnant, so that an almost complete restoration of liver mass is achieved. Cells of the renal tubule can be induced to proliferate by treatment with folic acid. Such tissues are sometimes called *conditional renewal populations*.

## 5.3  Classification of cell populations: II

Gilbert and Lajtha (1965) proposed a more complex subdivision of populations into *compartments* which have both structural and functional aspects. They distinguish, for example (Fig. 5.1):

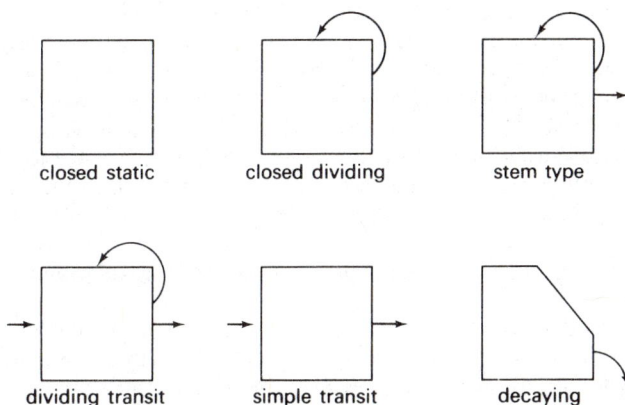

**Fig. 5.1** Population compartments expressed diagrammatically showing flux into the compartment (arrow entering) and out of it (arrow leaving). Arrows above the compartment denote daughter cells, born by division within the compartment *en passant*. The 'decaying' compartment is defined in the text but will not be discussed (redrawn, slightly modified, from Cleaver, 1967).

(a) *a closed static compartment*, in which there is no renewal of cells and a negligible loss, e.g. the neurone population of the central nervous system;

(b) *a decaying compartment* where again there is no renewal of cells, i.e. no input into the compartment after birth. This compartment is distinguished from (a) by the fact that cells are lost over a period of time. It is exemplified by germ cells in the adult ovary and may be approximated by a regressing tumour after radiotherapy or other cytotoxic treatment;

(c) *a stem cell compartment* with proliferation and output but no input. A population (such as the basal layer of the epidermis) which maintains itself by continuous proliferation and at the same time supplies cells continuously to other compartments, i.e. the layers above it;

(d) *a simple transit compartment* in which cells have ceased to proliferate. This compartment has input and output but no proliferation. The reticulocyte of the bone marrow or the cells in the upper portion of the small bowel crypts are typical examples;

(e) *a dividing transit compartment* such as may be found in the bone marrow and the proliferating portion of the small bowel crypt, where there is input, further proliferation, and eventually output of cells; and finally the rather special entity;

(f) *a closed dividing compartment*, which is one with no input or output of cells. Cells proliferate and are retained, leading to an increase in the size of the compartment. Examples are proliferating tissue cultures, regenerating tissue (e.g. liver after partial removal), and tumours in which we make the provisional assumption that there is no loss of cells.

In growing normal tissues and in tumours, cell production exceeds cell loss and these are therefore *expanding* populations. Adult renewing populations maintain a constant size, with cell production equalling cell loss; such populations are said to be in a *steady state*. These real cell populations consist of several of the above compartments linked together so that they feed sequentially one into another. The simplest type of renewal system has two compartments (Fig. 5.2*a*), namely, a stem cell compartment and differentiated cell compartment from which cells leave to die. In most renewal systems cells pass through an intermediate maturation phase during which proliferative activity is lost and characteristics of the mature cell type appear. This maturation compartment is thus a simple transit compartment and leads to the arrangement shown in Fig. 5.2*b*. In addition, renewal systems such as the bone marrow have one or more dividing transit compartments leading to the more complex compartment model shown in Fig. 5.2*c*.

The existence of potentially fertile but resting $G_0$ cells has already been mentioned. There is some evidence that, even in renewal systems, $G_0$ cells are present and, in certain conditions, can proliferate and increase the rate of influx into dividing compartments, such as the stem cell compartment or a dividing transit compartment, as shown in Fig. 5.2*d*. Such $G_0$ reserves are thought to be present in bone marrow and epidermal stem cell populations.

The reader should note that we have used two kinds of classification in the paragraphs above. First, there is the segregation of *cells* into proliferating and non-proliferating classes. Secondly, there is the segregation of *tissues* into compartments of different kinds. These classifications cut across each other; thus, for example, we may have proliferating cells in stem cell compartments and also in dividing transit compartments. In general we speak of *cells* in *phases*, and *tissues* in *compartments*.

### 5.4  A simple model of a renewal cell population

In this section we describe briefly a cell kinetic model of a simple renewal popu-

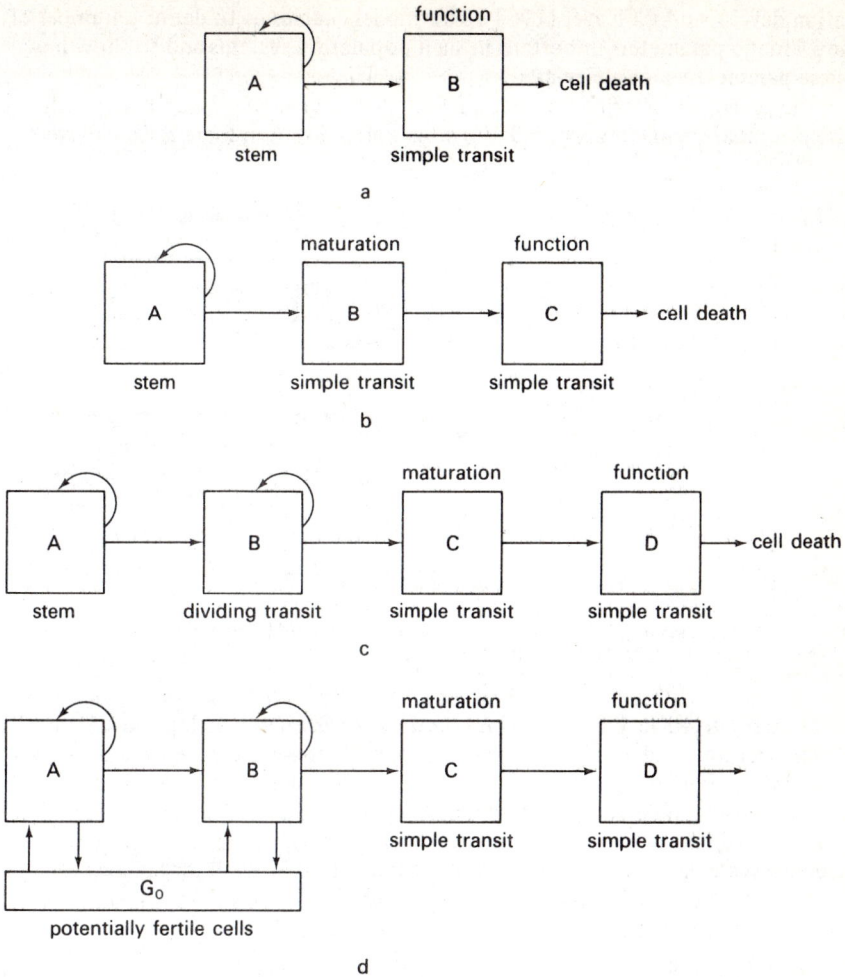

**Fig. 5.2** (*a*) The simplest type of renewal system – a self-renewing stem cell compartment (A) feeding a simple transit functional compartment (B); (*b*) a simple three-compartment renewal system comprising a stem cell compartment (A) maintaining itself and feeding a simple transit maturation compartment (B) which in turn feeds a functional compartment (C); (*c*) a four-compartment renewal system. Here a stem cell compartment (A) feeds a dividing transit compartment (B); both maintain themselves as well. Cells which enter the maturation compartment (C) lose the ability to divide before they finally enter the functional compartment (D); (*d*) the four compartment model shown in (C) altered by adding a $G_0$ phase, which consists of potentially fertile cells not involved in division but capable of entering the cycle and feeding cells into both the stem cell compartment (A), and the dividing transit compartment (B) (redrawn from Cleaver, 1967).

lation developed by Cleaver (1967). This model enables us to define a number of new kinetic parameters important in such population systems and to show how these parameters are interrelated.

The model we describe is a renewal system in a *dynamic steady state*, with three compartments or subpopulations. As Fig. 5.3 shows, there is a small *stem*

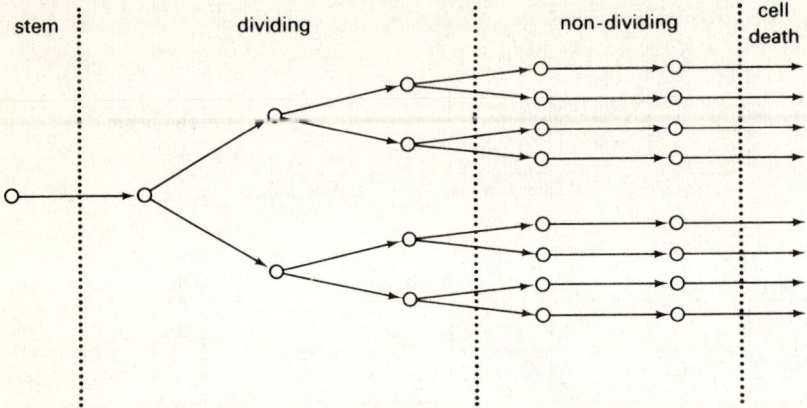

**Fig. 5.3** A tree diagram of the three-compartment model population developed by Cleaver (1967). The relationships between the compartments are expressed in Equations 5.1 and 5.2; see the text.

*cell compartment* feeding a *dividing transit compartment* which in turn feeds a *simple transit compartment*. The latter may be supposed to be a maturation or a functional compartment. While they are in the dividing transit compartment cells are assumed to undergo three division cycles. Rather arbitrarily we further assume: (a) that the duration of the cell cycle ($t_C$) is constant throughout the renewal system, and (b) that there is no cell death other than that indicated.

We symbolize the number of cell divisions as $m$ (in this instance $m = 3$): $m$ is related to the *fluxes* into and out of the compartments thus:

$$k_{in} = 2^m k_{out} = 2^3 k_{out} \tag{5.1}$$

where k stands for the *rate of flux*, i.e. *input rate* into a compartment *or output rate* from it depending on the subscript. Note that we cannot properly speak of transit divisions in the case of the stem cell compartment. There is no input; stem cells renew themselves. There is however, a *turnover time*, i.e. the time taken to replace every cell of the compartment by a new one.

Solving Equation 5.1 for $m$, in terms of common logarithms, we have

$$m = 3 \cdot 3 (\log_{10}) \left[ \frac{k_{out}}{k_{in}} \right]. \tag{5.2}$$

Figure 5.3 and Equation 5.2 enable us to establish the relationships shown in Table 5.1. The non-dividing compartment is *simple transit* and has no transit divisions.

The birth rate $k_B$ may be calculated from the reciprocal of the cell cycle time, provided all cells are in cycle, and, in the stem cell compartment of Fig. 5.3, $k_B$

**Table 5.1** Relationships between population compartments and their population kinetics

| Parameter | stem | dividing transit | non-dividing (maturation) |
|---|---|---|---|
| | | Compartment | |
| Size | 1 | 7 | 16 |
| Cell cycle time, $t_C$ (hr) | $t_C$ | $t_C$ | – |
| Number of cell cycles ($m$) during transit | – | 3 | – |
| Cell birth-rate, $k_B$ (cells/cell/hr) | $1/t_C$ | $7/t_C$ | – |
| Input, $k_{in}$ (cells/cell/hr) | 0 | $1/t_C$ | $8/t_C$ |
| Output, $k_{out}$ (cells/cell/hr) | $1/t_C$ | $8/t_C$ | $8/t_C$ |
| Transit time, $t$ (hr) | – | $3t_C$ | $2t_C$ |
| Turnover time* | $t_C$ | $7t_C/8$ | $2t_C$ |

\* When the stem cell compartment cannot be distinguished from the dividing compartment the turnover time, in this combination of compartments, is $t_C$. Turnover time is defined as the time taken for all cells in a given class to be replaced.

will be $1/t_C$, while in the dividing compartment $k_B$ will be $7(1/t_C)$ or $7/t_C$. The non-dividing compartment, of course, has no birth rate.

The stem cell compartment by definition has no influx, and of course the influx into the dividing compartment will equal the efflux from the stem cell compartment. The efflux from the stem cell compartment will equal the birth-rate, $1/t_C$ which will be the influx into the dividing compartment. The efflux from the dividing compartment will equal the sum of the birth rates in all preceding compartments, $1/t_C + 7/t_C$, or $8/t_C$. Now since the non-dividing compartment is simple transit, the influx will equal the efflux, and since there is no birth rate, this will be equal to the efflux from the dividing compartment, or $8/t_C$.

The stem cell compartment must have special arrangements to maintain its size. For this to occur, on average every stem cell division provides the subsequent dividing compartment with one cell, and one remains behind to maintain the stem cell compartment. For example, in a simple population, stem cells ($P_s$) become non-dividing functional cells ($Q_f$); symbolically

$$P_s \rightarrow Q_f$$

where subscripts are used to indicate that the proliferating cells in question are in the stem cell compartment, and the non-proliferating cells are functional. It is necessary to make these distinctions because the P–Q classification overlaps the compartment type classification. One way of maintaining the stem cell compartment size would be by asymmetrical division

$$P_s \rightarrow P_s + Q_f.$$

The concept of asymmetrical division has met with criticism on various counts (Osgood, 1957; Cairnie, Lamerton and Steel, 1965). An alternative mechanism might be

$$P_s \rightarrow P_s + P_s$$

and

$$P_s \rightarrow Q_f + Q_f$$

which avoids the problem of asymmetrical division and provides a means whereby equal numbers of $P_s$ and $Q_f$ cells could be produced.

We have already noted that a stem cell compartment cannot be said to have a true transit time in the usual sense of the term, and so a value cannot be quoted. In the proliferative compartment there are three transit divisions, and consequently, with a cell cycle time $t_C$, the transit time will be $3t_C$. To calculate values for the transit time through the maturation compartment we note that the system is in a steady state and cell death does not occur in it. This being so, and given that influx equals efflux,

$$\sum_j k_{ji} = \sum_j k_{ij}$$

where $\sum_j k_{ji}$ simply indicates the sum of all the inputs into the $i$th compartment,

and $\sum_j k_{ij}$ the total efflux. Hence, because we have a steady state the transit

time $t_i$ is equal to the ratio of the compartment size $n_i$ to the total influx (or efflux, in the steady state); thus, on average,

$$t_i = n_i / \sum_j k_{ij}.$$

Applying this to the maturation compartment we have $t_i = 16/8/t_C$ which simplifies to $2t_C$.

For the stem cell compartment, the turnover time will equal the cell cycle time under the assumptions made above. In the proliferative compartment the turnover time is almost the same as the cell cycle time, since most cell production takes place within this compartment. If, however, there were a greater input from the stem cell compartment, then the turnover time would be *less* than the cell cycle time. In the non-dividing maturation compartment the turnover time will be equal to the transit time, i.e. $2t_C$.

## 5.5  A simple model of an expanding cell population

Bresciani (1968) has proposed a classification of tissues similar to those we have discussed above, together with a set of equations by which the effect of a change in one or more parameters can be deduced. The essential features of his system are the usual four-phase cycle and a subdivision of tissue populations into progenitor, expanding–maturing, and non-proliferative compartments, in that sequence. The progenitor compartment (P) is self-maintaining, the expanding–maturing compartment (E, M) — cf. the dividing transit compartment of earlier systems — may possibly be absent, and the non-proliferative (Q) compartment contains cells in various states such as $G_0$ and an irreversibly differentiated state which Bresciani calls z-cells.

His distinctive contribution to cell population analysis is the 'distribution ratio' (d-ratio) which we have already met in Section 2.3.2, p. 13. There we noted that the d-ratio is $\overleftarrow{\eta}/\overrightarrow{\eta}$ where $\overleftarrow{\eta}$ is the fraction of newborn cells which recycle in a proliferative compartment, progenitor or expanding, and $\vec{\eta}$ is the fraction which decycles and passes on to the next compartment in the sequence. In the normal expanding–maturing compartment a cell proceeds through one or more successive divisions following an irreversible differentiation pathway. Therefore, although proliferation may occur, $\overleftarrow{\eta} = 0$ and $\vec{\eta} = 1$. But in pathological (e.g. tumour) tissue an increasing pool of recycling cells could form in the P-compartment or the E, M-compartment, or both. Then $\overleftarrow{\eta}_P > 0 \cdot 5$ or $\overleftarrow{\eta}_{E,M} > 0$, or both.

We distinguish two kinds of expanding population; first, that in which expansion is a temporary phase between one steady state and another more populous steady state; and secondly, that in which expansion is (ideally) unlimited, as in tumours. In the first kind $\overleftarrow{\eta}/\vec{\eta} = 1$, and in the second $\overleftarrow{\eta}/\vec{\eta} > 1$ because $\overleftarrow{\eta} > 0 \cdot 5$ and therefore proliferative cells are being retained in some compartment.

Consider a population consisting of the two fundamental compartments, the progenitor or stem cell compartment and the differentiating, functional, non-proliferative compartment; in short the P and the Q subpopulations. Let the d-ratio be unity. This population is in a steady dynamic state (Fig. 5.4). Now

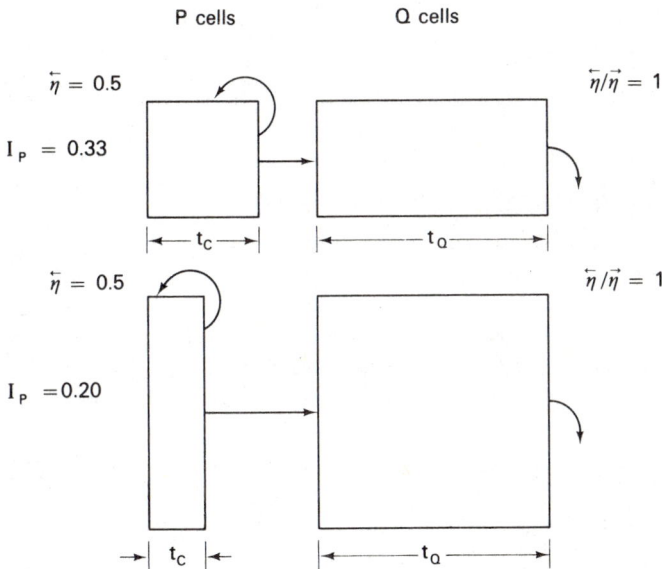

**Fig. 5.4** This diagram, redrawn from Bresciani (1968), shows in its upper part a population consisting of a stem cell compartment (P) and a non-proliferative compartment (Q). Of each pair of daughter cells produced in P one proceeds to Q and the other remains in P to start a new cycle; therefore $\overleftarrow{\eta} = 0 \cdot 5$, $\overleftarrow{\eta}/\vec{\eta} = 1$ and the whole system is in a steady state. In the lower part of the diagram we suppose that the rate of cell proliferation has doubled throughout P (i.e. $t_C$ is halved). This leads to a doubling in the size of Q due to the doubled rate of influx from P.

suppose that the duration of the cell cycle, $t_C$, is reduced to half in some fraction $\lambda$ of the progenitor compartment. Of course $\lambda$ may be unity, in which case *all* of the stem cells are replicating twice as fast as before and the flux from the progenitor to the differentiating compartment is doubled. The final size of the whole population, or its new steady state level, assuming $\overleftarrow{\eta}/\overrightarrow{\eta} = 1$, may be calculated from

$$N = n_P + [(t_C)_a/(t_C)_b] \, n_Q \qquad (5.3)$$

where $N$ is the new steady state level, $n_P$ is the number of progenitor (stem) cells, $n_Q$ is the number of non-proliferating cells, and the subscripts $a$ and $b$ refer to the value of $t_C$ before and after the moment of acceleration.

For example, if $t_C$ were halved in $\lambda = 10$ per cent of the progenitor cells, and the growth fraction at that time were known to be $I_S = 0.33$ we could write

$$N' = 0.1N = 0.1(n_P + 2n_Q); \quad N' = \lambda N. \qquad (5.4)$$

From the growth fraction we have

$$n_P/(n_P + n_Q) = 0.33,$$

therefore

$$n_Q = 2.03 \, n_P.$$

Rewriting 5.4 with this simplification,

$$0.1 \, N = 0.1 \, [n_P + 2(2.03)n_P] = 0.506 \, n_P. \qquad (5.5)$$

The new steady state level is therefore

$$0.9 \, N_a + 0.1 \, N_b = 2.727 \, n_P + 0.506 \, n_P = 3.233 \, n_P$$

that is to say, the whole system will have enlarged by 6.7 per cent, and the new $I_S = 0.304$. We have expressed the increase in terms of $n_P$ because $n_P$ is easily found if the growth fraction and the population size are known. But in fact the expansion is transferred to the Q compartment, due to the accelerated flux from P. In due course accelerated cell loss leads to the steady state.

Should the residence time in Q be prolonged a similar calculation may be made, since

$$N = n_P + [(t_Q)_a/(t_Q)_b] \, n_Q. \qquad (5.6)$$

Similar arguments apply to three-compartment populations. For example

$$N = n_P + [(t_C)_a/(t_C)_b] \, (n_{E,M} + n_Q). \qquad (5.7)$$

A profound change in the character of the population occurs if $\overleftarrow{\eta}/\overrightarrow{\eta} \neq 1$. In a steady state one cell in each pair of daughter cells produced by mitosis returns to the cycle, and therefore $\overleftarrow{\eta} = 0.5$. The equation characteristic of this state is

$$n_{P(t)} = n_{P(o)}(2\overleftarrow{\eta})^{t/t_C} \qquad (5.8)$$

from which it is evident that after an interval of time equal to the cell cycle time, $n_{P(t)}$ reverts to its starting value $n_{P(o)}$. But if $\overleftarrow{\eta} > 0.5$ then $n_{P(t)} > n_{P(o)}$. In general, as Bresciani shows and illustrates as in Fig. 5.5,

$$n_{P(t)} = n_{P(o)} \exp[\ln (2\overleftarrow{\eta}) t/t_C] \qquad (5.9)$$

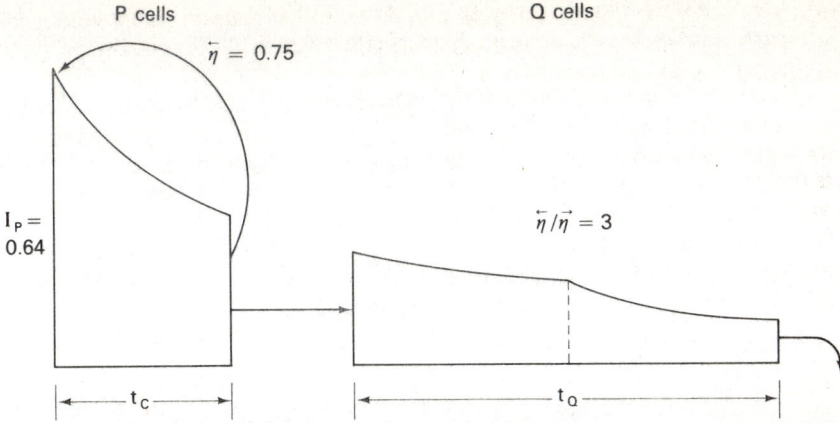

Fig. 5.5 This diagram, also redrawn from Bresciani (1968), shows what happens when the proportion of daughter cells remaining in P is greater than half, i.e. when $\overleftarrow{\eta} > 0.5$; here we have depicted $\overleftarrow{\eta} = 0.75$. The steady state no longer obtains. Instead the population expands, theoretically without limit. In this state $\overleftarrow{\eta}/\overrightarrow{\eta} > 1$; here the ratio is 3. Note that the rectangular age distribution is no longer appropriate; instead we have the negative exponential distribution discussed in Chapter 2.

and the size of the Q-compartment at time $t$ is

$$n_{Q(t)} = \sum_{t-t_Q}^{t} [t_C(\overrightarrow{\eta}/\overleftarrow{\eta})n_{P(o)} \exp (\ln 2\overleftarrow{\eta}) t/t_C]. \tag{5.10}$$

For further details the reader should consult Bresciani (1968).

These are simple models of compartmental population kinetics. In real situations it is usually not possible to define a population so exactly. This is particularly true in *tumour* populations, in which spatial, morphological, and functional compartments are difficult and often impossible to define. This leads to serious limitations in many studies. For example, it is almost certain that tumours have within them a stem-cell compartment, but at present little can be said about it. This is a limitation because it seems likely that tumours which have been sublethally affected by treatment may regrow from a small stem-cell compartment. It could be of clinical value to know more about the kinetics of such a compartment.

## Summary

Cells aggregate into populations with distinctive characteristics; some populations (e.g. neurones) may be static, some (e.g. epidermis, bone marrow, intestinal mucosa) may be in a state of steady continuous renewal, and some (e.g. tumours) are typically expanding populations.

Populations, in their turn, may be subdivided into compartments. Different schemes have been proposed but most have the following features in common: a stem-cell compartment characterized by proliferation, a maturing compartment where cells differentiate but may also undergo further division, and a functional compartment where cells perform their specific roles until they finally succumb to the process of ageing.

The size of a compartment and the characteristic age distribution of its cells are determined largely by the proportion of cells which leave the proliferative cycle (i.e., decycle), and the consequent proportion which continue to cycle (i.e., recycle). Changes in these proportions also affect the characteristics of subsequent compartments.

Two model populations are described, one a simple renewal type and the other an expanding type; equations are given by which the effects of changes in one or more population parameters can be deduced.

## References

BRESCIANI, F. (1968) Cell proliferation in cancer. *Europ. J. Cancer*, 4, 343.

CAIRNIE, A. B., LAMERTON, L. F. and STEEL, G. G. (1965) Cell proliferation studies in the intestinal epithelium of the rat. II Theoretical aspects. *Exp. Cell Res.*, 39, 539.

CLEAVER, J. E. (1967) *Thymidine metabolism and cell kinetics*. North Holland, Amsterdam.

GILBERT, C. W. and LAJTHA, L. G. (1965) In *Cellular Radiation Biology*, p. 474. The University of Texas, M. D. Anderson Hospital and Tumour Institute.

LEBLOND, C. P., MESSIER, R. B. and KOPRIWA, B. (1959) Thymidine-$^3$H as a tool for the investigation of the renewal of cell populations. *Lab. Invest.*, 8, 296.

OSGOOD, E. E. (1957) A unifying concept of the aetiology of the leukaemias, lymphomas and cancers. *J. Nat. Cancer Inst.*, 18, 155.

# 6 The dynamics of growth

The over-all growth rate of a proliferating population is the resultant of several factors, most of which we have discussed in Chapters 3 to 5. We are now in a position to consider methods of calculating: (a) the *overall growth rate*, (b) the proportion of cells actually in cycle, i.e. the *growth fraction*, and (c) the rate at which cells are lost from the population. Factors (b) and (c) are the most important of the factors which govern the over-all growth rate (a). We shall not consider factors such as haemorrhage, necrosis, or cyst formation though these are prominent features of neoplasms. In the latter part of this chapter we shall discuss growth curves briefly.

## 6.1 Growth parameters

The growth rate ($k_G$) and over-all volume doubling time ($t_D$) are two parameters of particular relevance to the study of tumour cell kinetics, though they apply to any expanding cell population. In theory, $k_G$ and $t_D$ are found directly by counting the increase in the number of *tumour cells* with the passage of time. This, of course, is generally impossible. The growth rate of solid tumours is usually deduced in practice (a) from serial caliper measurements of external tumour volume, or (b) from serial measurements of pulmonary or osseous metastases as seen on radiographs. This latter method is particularly relevant to the study of *human* cancers where direct measurement is generally not possible. In studies of experimental tumours implanted subcutaneously in animals, a calibration graph relating external measurements to tumour weight can be constructed by removing and weighing a series of externally measured tumours over a period of time.

None of these methods is entirely satisfactory, for a number of reasons. Caliper measurements are, of necessity, limited to use with subcutaneous tumours and inaccuracies are introduced by the need to make measurements through a layer of skin. Further, calculation of volume from these measurements is usually based on the assumption that tumours are segments of spheres, or of ellipsoids, which they rarely are. In human studies a further difficulty arises from the short period that most tumours are available for observation before treatment is begun. Whatever the technique there are still factors other than increase in cell number which may contribute to an increase in tumour mass: cellular hypertrophy, infiltration of leucocytes, secretion of fluid, formation of collagen, and haemorrhage are all common factors. Clearly, therefore, although morphometric techniques are of use in separating the contributions made by such factors, the measurements of growth rate and doubling time are generally subject to considerable inaccuracies.

The over-all growth rate ($k_G$) and the volume doubling time ($t_D$) are found as follows. Serial measurements of population mass (or volume) are first plotted on arithmetical or semi-logarithmic graph paper. This provides the *growth curve*, about which we shall have more to say presently. The growth rate at any particular point $t$ in time is then found by drawing a tangent to the growth curve at that point. The slope of the tangent is equivalent to $k_G$. To find the over-all volume doubling time at $t$ in an exponentially growing population we note that, generally,

$$V_t = V_0 \exp(k_G t) \tag{6.1}$$

and therefore at time $t$

$$V_t/V_0 = 2 = \exp[k_G(t)\, t_D(t)]$$

so that

$$t_D(t) = \ln 2/k_G(t). \tag{cf. 2.14}$$

This equation for $t_D$ is written explicitly as $t_D(t)$ to imply that the value of the over-all volume doubling time is not, in reality, a constant quantity. Similarly, $k_G(t)$ is meant to imply that the value measured by the tangent is not constant either. Both values are true only for the time point $t$.

## 6.2 Techniques for estimating the growth fraction

Two methods are available for estimating the growth fraction by means of the FLM technique. The first of these requires that the pulse labelling index ($I_S$) is known and that two peaks are obtained on the FLM curve. As we saw in Chapter 2 the growth fraction is

$$I_P = \frac{\text{Number of cycling cells }(N_C)}{\text{Total cells }(N)}. \tag{cf. 2.1}$$

This equation can be rewritten as

$$I_P = (N_S/N)/(N_S/N_C) \tag{6.2}$$

where $N_S$ = number of cells in S, $(N_S/N)$ is the *actual* counted labelling index ($I_S$), and $(N_S/N_C)$, often referred to as the *theoretical* labelling index, is the index which would be expected from the FLM data if *all* of the cells were in cycle. If steady state conditions apply in the population under study, and the population has a rectangular phase distribution diagram, then the number of cells in a given phase of the cycle is directly proportional to the duration of that phase. Thus $(t_S/t_C) \equiv (N_S/N_C)$. Both $t_S$ and $t_C$ can be obtained from the FLM curve and thus $I_P$ can be calculated from Equation 2.6. In an exponentially growing population we have

$$(N_S/N_C) = [\exp(t_S \ln 2/t_C) - 1]\,[\exp(t_2 \ln 2/t_C)] \tag{cf. 2.6}$$

The second method for estimating $I_P$ from FLM curve data is mainly used when a second peak is not obtained and so an accurate estimate of $t_C$ cannot be made. In this case the value of $N_S/N_C$ can be estimated as follows. We have seen (Section 4.5) that successive waves of labelled mitoses damp out, due to variations in the duration of the cell cycle phases. The fraction of labelled mitoses thus tends to an equilibrium value, which is equal to $N_S/N_C$ when the labelled cohort is completely desynchronized. In practice one finds this value by counting labelled mitoses in samples taken sufficiently long after the initial labelling pulse for damping to be complete.

The reason why the eventual equilibrium proportion of labelled mitoses should equal $N_S/N_C$ is as follows. Cells in mitosis are by definition in cycle and so by looking at mitoses one is excluding non-cycling cells. When the labelled cohort of cells has become desynchronized the proportion of cells labelled in each of the cell cycle phases and in the cell cycle as a whole will be the same. Thus, if 30 per cent of mitoses are labelled it follows that 30 per cent of all cycling cells will be labelled and we have our value for ($N_S/N_C$). The growth fraction can then be calculated from Equation 6.2.

Unfortunately, this method is beset by a number of technical difficulties. For example, during the period which one must allow for adequate damping, the labelled cells go through a number of division cycles. At each division the quantity of radioactive thymidine in them is halved. Thus the label may be diluted to such an extent that mitoses which should be labelled are not recognized as such. Conversely, if cells die, as is likely during long-term experiments, the label they release can be taken up and re-used by cells not in the original labelled cohort. In this way mitoses come to be labelled which should not be. The extent of these sources of error depends on the manner in which the experiment is carried out and the nature of the population under study.

The continuous or repeated labelling technique (Section 4.6) also enables an estimate of growth fraction to be made. In such experiments a plateau value for the fraction of cells labelled is reached in a time equal to $t_{G2} + t_M + t_{G1}$ (Fig. 4.9). Ideally all cycling cells will have become labelled by this time and thus the fraction of labelled cells at this interval should be a measure of the growth fraction. A limitation of this technique is that it tends to overestimate the growth fraction in populations where there is an appreciable transition of cells from the cycling to the non-cycling state. The error caused will increase with the duration of the labelling period and also with the life span that cells have after leaving the cycle. On the other hand, if labelled cells die, in cycle, then $I_P$ could be underestimated).

## 6.3 Cell loss

The rate of cell loss is as important a factor as the rate of cell production in determining the dynamic state of a cell population. An *accurate* assessment of the rate of cell loss is difficult to obtain, even in the experimental animal. We must necessarily work with approximations.

There are two fundamentally different ways of assessing cell loss. The first is an indirect method, in which cell loss is estimated from the discrepancy between (a) the apparent cell cycle time, $t_{C(a)}$, as determined by a stathmokinetic study, and (b) the over-all volume doubling time, $t_D$, which we discussed above (Section

6.1). The cell loss factor ($\phi$) is obtained, as we saw in Chapter 2, from

$$\phi = 1 - (t_{C(a)}/t_D). \tag{cf. 2.15}$$

The second, more direct, approach to assessing the rate of cell loss entails monitoring the loss of radioactivity from a cell population after it has been given a pulse of a radioactively labelled compound which is incorporated into DNA. Samples of tissue are taken at intervals subsequently. As labelled cells are lost, the total tumour radioactivity will decrease and the rate of this decrease will give a measure of the cell loss rate. Initially $^3$H-thymidine was used to label cells, but it was found that cells, on dying, released their $^3$H-TdR which was then picked up and re-used by surrounding healthy cells. To meet this technical problem compounds have subsequently been developed which are much less readily re-used than is $^3$H-TdR. One such compound is $^{131}$I-labelled 5-iodo-2-deoxyuridine. Though the re-use of this compound is less, the problem is not entirely solved, particularly in situations where cells are dying in large numbers. Moreover, compounds such as $^{131}$I-labelled 5-iodo-2-deoxyuridine are toxic, both chemically and radiologically; great care must therefore be taken in selecting a suitable dose. In most cases the radioactivity in each tissue sample is estimated by scanning techniques which require prior chemical extractions of DNA. This tends to prolong the procedure. It may be possible in some cases to overcome this difficulty by using an external scanning technique to estimate the total radioactivity in a localized superficial tumour transplant. This would have the additional merit of not perturbing the tumour in question, a risk which is always present when methods which require multiple biopsies are used.

## 6.4  Growth equations

The number of individuals, $N$, in an expanding population increases according to some function of time $f : t \to N$. If we assume: (a) that in any small interval of time, $\Delta t$, a new individual appears in the population with constant probability $\lambda \Delta t$; (b) that the rate of change in the size of the population depends *only on the state* at time $t$; and (c) that it is permissible to approximate a real *discrete* growth function by a *continuous* differentiable function, we can then derive the exponential growth law

$$M_t = M_0 \exp k_G t \tag{6.3)(cf. 6.1}$$

where $M$ is the *mass* of the population, disregarding individuals as such, and $k_G$ is the growth rate. Note, in passing, that we assume $M_t \propto V_t$.

The exponential growth curve matches empirical graphs rather poorly, but it is sometimes useful to approximate short segments of empirical curves by an exponential formula.

Before we seek a more realistic growth function we must introduce an important concept, namely the *specific growth rate*, which we shall symbolize as SGR. This, essentially, is the first derivative of the growth function. It relates the *rate* of growth to the *size* of the population at any given time. For example, in the exponential case we have

$$dM_t/dt = kM,$$

or

$$M^{-1}(dM_t/dt) = \text{constant}. \tag{6.4}$$

This may be interpreted as meaning that the larger the mass of the population the faster the rate of growth, a conclusion that runs counter to our intuition.

The advantages of the exponential approximation are: (a) information is needed about only two parameters ($M_0$, k); and (b) the mathematics to which it gives rise is simple.

Suppose we take the more plausible view that the SGR *decreases* exponentially as the population increases in size (Fig. 6.1*a*). This may be written

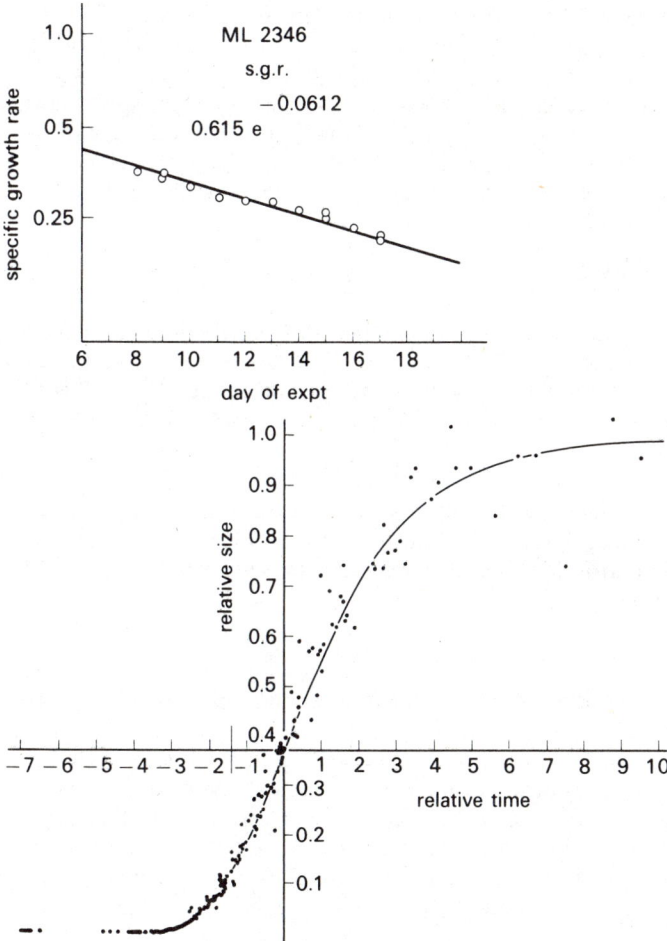

Fig. 6.1 The upper graph (*a*) shows an exponential decrease in the value of the Gompertzian specific growth rate; the values give a linear graph when plotted, as here, on a semi-logarithmic grid. The tumour in question is a metastasizing lymphoma (ML) kindly supplied by Dr. Richard Carter. The lower graph (*b*) is a Gompertz curve fitted to growth data derived from a number of experimental animal tumours. It is reproduced, with added legends, by courtesy of Dr. Anna K. Laird and the editor of the *British Journal of Cancer*.

$$M^{-1}(dM_t/dt) = k \exp(-rt) \tag{6.5}$$

where r is a constant that governs the decrease in growth potential. A suitable integration of **6.5** leads to the Gompertz growth function

$$M_t = M_0 \exp\{A/r\,[1 - \exp(-rt)]\} \tag{6.6}$$

where A is a growth rate constant.

This function may be written in various forms. The form we use (with a minor change in notation) is that which Laird (1964, 1966) in particular has applied to both normal and neoplastic growth. Figure 6.1*b* shows the realistic sigmoid shape of the curve, and the excellent fit to a number of experimental neoplasms. We shall see another excellent fit to one of our own neoplasms in Chapter 7.

The disadvantage inherent in the Gompertz curve is that it requires information about three parameters ($M_0$, A, r) and is therefore more difficult to construct in a given case, especially in the human situation. Laird (1964) has drawn attention to some properties of the Gompertz growth function which are of particular interest in the study of neoplasms. For instance, if the period of growth is prolonged the mass of the neoplasm does not increase indefinitely, as the exponential function predicts, but tends to a limit:

$$\lim_{t \to \infty} M_t = M_0 \exp(A/r). \tag{6.7}$$

Though this limit may not be attained in practice, its existence in theory conforms to current thinking about the behaviour of neoplasms. Secondly, the number of times a given neoplasm doubles its mass during its period of growth is

$$n = (A/r)\ln 2. \tag{6.8}$$

These equations bring out the importance of the ratio A/r in determining at least some of the growth characteristics of a particular neoplasm. Indeed, it is tempting to suppose that A/r, and the value of r in particular, are numerical manifestations of an intrinsic process whereby growth is regulated.

## Summary

The most immediate measure of growth in a proliferating population is the over-all volume doubling time. This is ordinarily assumed to refer to the proliferating cells only, though of course other cells and tissues (e.g. the vasculature) always constitute a significant proportion of the over-all volume. The doubling time is estimated from growth curves based on direct caliper measurements of experimental transplantable subcutaneous tumours in the laboratory animal and from serial X-ray films of tumours in man. Thus, the overall volume doubling time, $t_D(t)$, and the growth rate constant; $k_G(t)$, may be established for that particular time point, $t$.

Over-all growth is a resultant of (a) the proportion of the cell population actively proliferating (the growth fraction), and (b) the rate at which cells are lost. This chapter presents methods by which these growth parameters are determined.

The simplest equation with which to characterize the growth of an expanding population is the exponential

$$V_t \text{ (or } M_t) = V_0 \text{ (or } M_0) \exp k_G t$$

where $V$ and $M$ stand for volume or mass of tissue respectively and $k_G$ is the growth rate constant. Unfortunately, with increasing $t$ this equation comes to describe an explosively rapid growth which is quite unrealistic. Since the growth curve is commonly observed to be S-shaped, with a tendency to form a plateau as the over-all growth rate diminishes (probably due mainly to a decreasing growth fraction), other more suitable but more complex growth equations have been proposed. The most realistic of these is the Gompertz equation, which may be regarded as combining exponential growth with exponentially increasing retardation of growth.

## References

LAIRD, A. K. (1964) Dynamics of tumour growth. *Br. J. Cancer*, **18**, 940.
LAIRD, A. K. (1966) Dynamics of embryonic growth. *Growth*, **30**, 263.

# 7 Worked examples of cytokinetic analysis

In this final chapter we shall gather up some loose ends and demonstrate analyses carried out on real tissues. We hope to highlight both the power and the limitations of the methods we have described in this booklet. The examples chosen are (a) the cell population kinetics of small bowel epithelium which illustrates the steady state, and (b) the kinetics of an experimental indigenous mouse sarcoma, and of a human cancer of the colon, both of which illustrate the features of expanding populations.

## 7.1 The small bowel renewal system

The continuously renewing cell population which makes up the small bowel mucosa has been extensively studied in recent years, not only for the light which it casts upon the organization of cell populations but also because of its clinical importance; small bowel epithelium is often affected during radiotherapy and cytotoxic chemotherapy because of its high rate of cell proliferation, and it has become necessary to understand the mechanisms of crypt repopulation after such insults.

The renewal unit in the small bowel epithelium is the *crypt of Lieberkühn*, shown diagrammatically in Fig. 7.1. Cell production occurs in the crypt, and from here cells migrate upwards on to the villi. Several kinds of cell are encountered in the system: the most common is the columnar epithelial cell, or enterocyte (Booth, 1968); the precursor cell, in this terminology, is an enteroblast. At the base of the crypt large granulated Paneth cells, of indeterminate function, are found, while higher in the crypts and on the villi are numerous mucus secreting goblet cells. There is good evidence that mature Paneth and goblet cells do not proliferate.

The simplest interpretation of small bowel epithelium is that it behaves as a two-compartment system. The region where cell production occurs, in the crypt, can be termed the *proliferative compartment*, which supplies cells to the *functional compartment*, i.e. the villi. We can map the spatial organization of these compartments by measuring the labelling index and mitotic index at each cell position within the crypt. This is done by counting labelled and mitotic cells at each cell position. These proliferative indices are then plotted against cell position. The method of construction of such a 'distribution curve' for the rat crypt is shown in Fig. 7.2.

The distribution curve for the labelling index in the rat jejunal crypt is shown

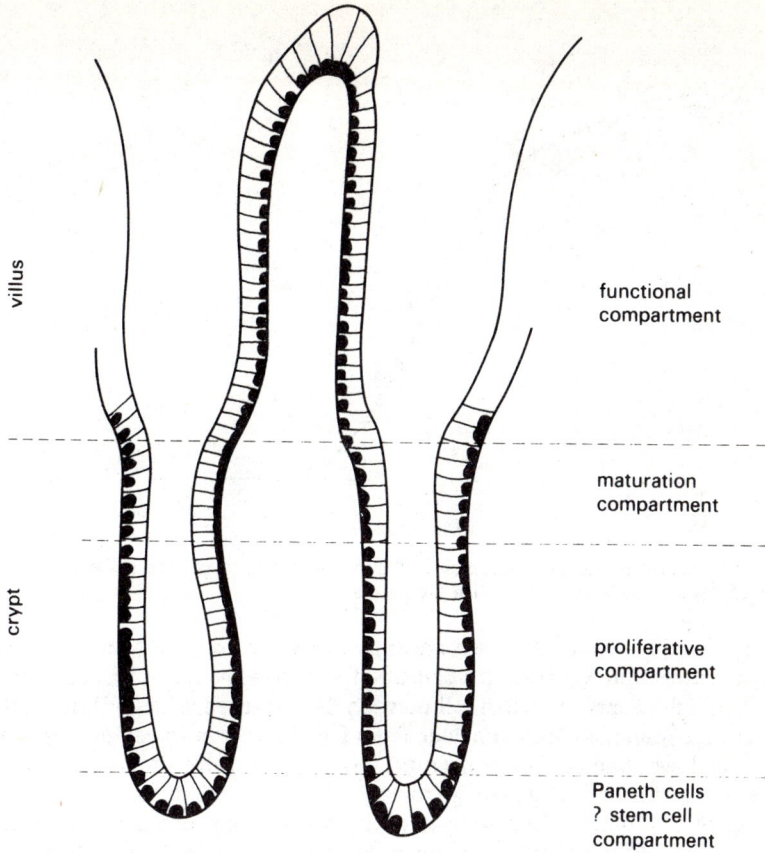

**Fig. 7.1** A diagrammatic representation of the kinetic compartments of the small bowel epithelium.

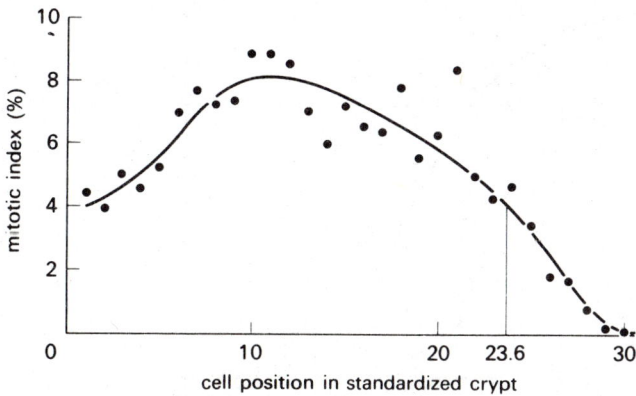

**Fig. 7.2** The method of construction of a crypt mitotic index curve in the rat jejunum. The mitotic index is measured at each cell position in sections of the crypt; thus by plotting the mitotic index as a function of cell position we can assess the relative rates of proliferation occurring at various levels in the crypt.

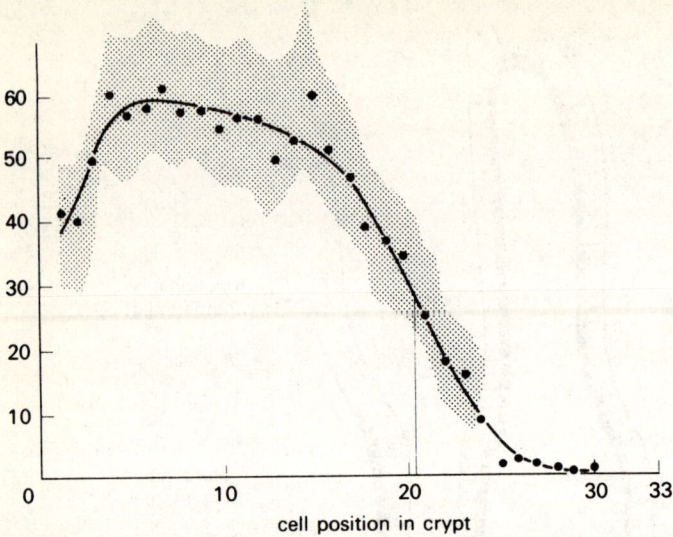

**Fig. 7.3** The labelling index distribution curve for the rat jejunal crypt. The shaded area indicates 95 per cent confidence limits for the points.

in Fig. 7.3. Comparatively low values are apparent in the lower cell positions, but the labelling index quickly rises to a plateau of almost 60 per cent. Higher in the crypt a sharp fall occurs and when cell position 24 is reached values of less than 10 per cent are found. We may conclude from Fig. 7.3 that most cell production occurs in the lower two-thirds of the crypt; this is the probable site of the proliferative compartment. But we can go further: in Fig. 7.4 the mitotic index is plotted against cell position and compared with the labelling index. As we might expect, mitoses are found further up the crypt than labelled cells, since, having

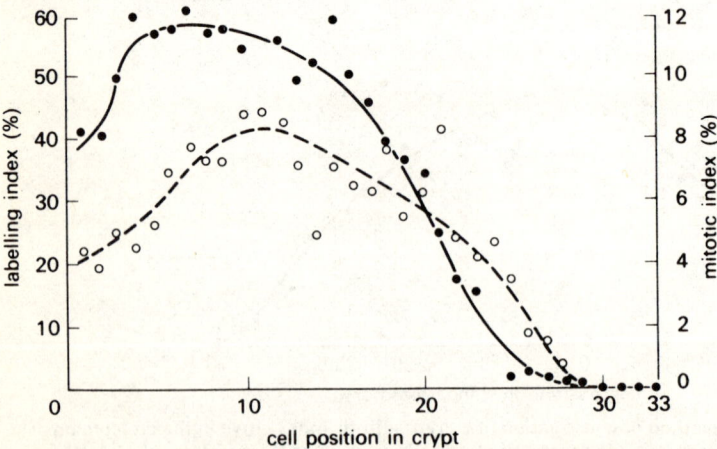

**Fig. 7.4** The crypt labelling index distribution curve compared with the crypt mitotic index distribution curve. Note that mitoses are found further up the crypt than labelled cells.

finished DNA synthesis, cells then go on to complete mitosis. Note that there are four cell positions in the uppermost part of the crypt which do not contain any labelled cells or mitoses. This immediately suggests a further compartment, which we term the *maturation compartment*, with the implication that cells continue the differentiation process during passage through this compartment, but do not divide again.

We have now delineated proliferative, maturation, and functional compartments: what of the stem-cell compartment? We could argue that since the cells at the crypt base have no input they form a stem-cell compartment. If this is so, the stem cells feed the proliferative compartment, which then becomes a *dividing transit compartment* as defined in Chapter 5 (p. 46). The maturation and functional compartments remain *simple transit compartments*. Cells are then lost into the intestinal lumen.

The cell cycle time is best estimated by the fraction of labelled mitoses (FLM) technique. For the rat, the mean cell cycle time is $t_C = 11 \cdot 3$ hours. In our simple model we assumed that $t_C$ was constant throughout the dividing compartment. However, FLM curves for different parts of the crypt showed that $t_C$ is about 15 hours in the base of the crypt and it can be seen that both $t_C$ and $t_S$ become shorter as we move up the crypt to position 8 (Fig. 7.5 and Table 7.1). Values are probably unchanged above this position. It appears that the cell cycle time in the basal portion of the crypt is (relatively) long. If the crypt base houses the stem cells, these long values could be confined to this compartment.

What is the size of the proliferative compartment? We have already seen from our graph of mitotic indices within the crypt that some cells at the top do not show any evidence of division. We can illustrate this further by calculating a growth fraction for the whole crypt, thus

$$I_P = (= f_G) = I_{S(ob)}/I_{S(th)} = 0 \cdot 62 \qquad (7.1)(\text{cf. } 6.2)$$

where $I_{S(ob)}$ is the observed labelling index, and $I_{S(th)}$ is the theoretical labelling index, i.e. that which would be obtained if *all cells were proliferating*. We therefore see that 62 per cent of crypt cells are actually in the cell cycle. But we cannot extrapolate and say that the remaining 38 per cent are erstwhile proliferative cells, now maturing, since the crypt also contains *non-proliferative* cells, i.e., Paneth cells and mature goblet cells.

The birth rate ($k_B$) for the proliferative compartment is

$$k_B = I_M/t_M \qquad (7.2)$$

expressed in cells per cell per hour; for the total birth-rate we multiply by the *number* of proliferative cells. By a separate study we found this to be 403 cells. The mitotic index is 3·46 per cent, and the mitotic duration 0·43 hours, and we have
$$k_B = (0 \cdot 0346/0 \cdot 43)403 = 32 \text{ cells/hour.}$$

We can confirm this value from the equation

$$k_B = I_P/t_C \qquad (7.3)(\text{cf. } 2.9)$$

again expressed in cells per cell per hour. The total birth-rate is
$$k_B = (0 \cdot 62/11 \cdot 3)650 = 36 \text{ cells/hour.}$$

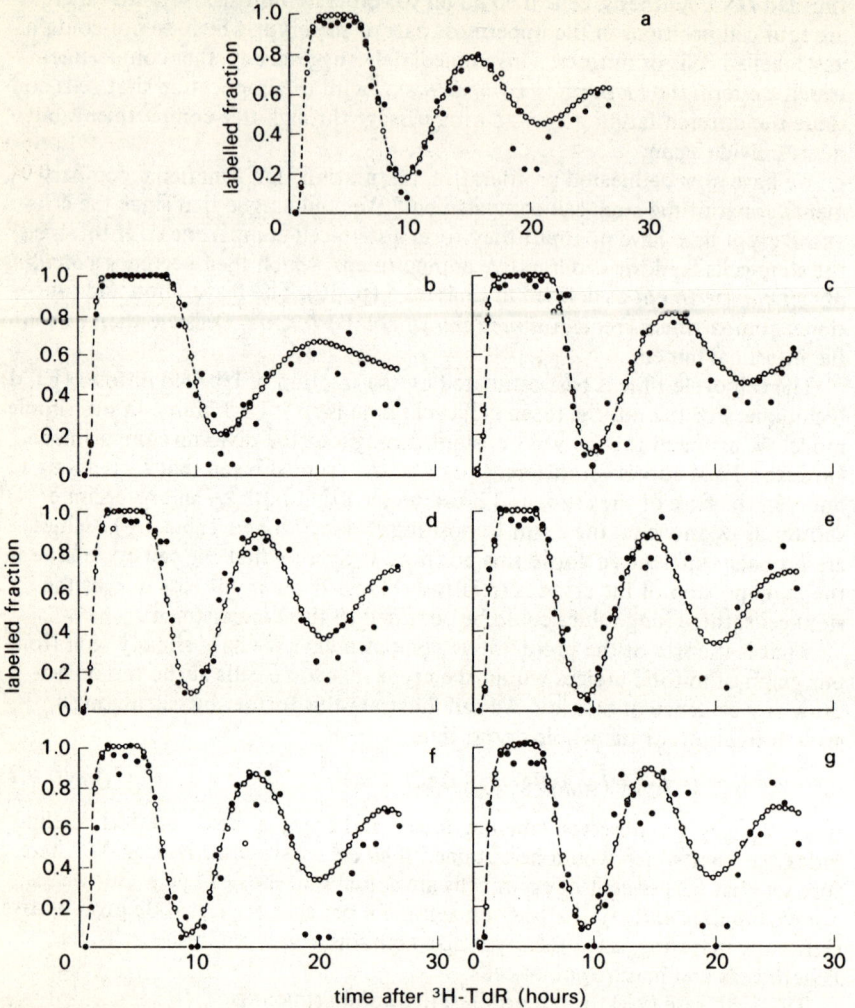

Fig. 7.5 Fraction of labelled mitoses (FLM) curves obtained from the rat jejunal crypt at different levels: (*a*) is for the whole crypt, (*b*) cell positions 1–4, (*c*) cell positions 5–8, (*d*) cell positions 9–12, (*e*) cell positions 13–16, (*f*) cell positions 17–20, and (*g*) cell positions above 20. The lines are fitted by the Gilbert (1972) technique. Note the damping in the curves for lower cell positions, indicating a wider spread of individual cell cycle times.

This birth-rate applies to all dividing compartments, and gives the efflux from these compartments. Since the subsequent compartments are of the simple transit kind the rate 35 cells/hour applies also to them.

Crypt cells probably move because of the proliferative activity in the cell positions below them. If this is so, the value obtained for the cumulative birth-rate at any cell position will also be equal to the speed at which cells are moving at that position, expressed in cell positions traversed per hour. The maximum reached in

**Table 7.1** Mean values (±SE) in hours for the duration of the cell cycle at various levels of the jejunal crypts of male Wistar rats.

|  | $t_C$ | $t_{G1}$ | $t_S$ | $t_{G2}$ |
|---|---|---|---|---|
| Cell positions 1−4 |  |  |  |  |
| Mean | 15·48 | 5·70 | 8·56 | 1·23 |
| SE | 0·24 | 0·24 | 0·10 | 0·04 |
| Cell positions 5−8 |  |  |  |  |
| Mean | 12·34 | 4·00 | 7·14 | 1·21 |
| SE | 0·14 | 0·16 | 0·08 | 0·04 |
| Cell positions 9−12 |  |  |  |  |
| Mean | 11·24 | 3·50 | 6·42 | 1·32 |
| SE | 0·14 | 0·20 | 0·14 | 0·06 |
| Cell positions 13−16 |  |  |  |  |
| Mean | 10·82 | 3·55 | 5·85 | 1·43 |
| SE | 0·14 | 0·19 | 0·13 | 0·06 |
| Cell positions 17−20 |  |  |  |  |
| Mean | 11·03 | 3·89 | 5·94 | 1·20 |
| SE | 0·15 | 0·20 | 0·14 | 0·06 |
| Cell positions above 20 |  |  |  |  |
| Mean | 10·66 | 3·24 | 5·96 | 1·46 |
| SE | 0·12 | 0·17 | 0·12 | 0·06 |
| Whole crypt |  |  |  |  |
| Mean | 11·32 | 3·52 | 6·49 | 1·30 |
| SE | 0·14 | 0·16 | 0·10 | 0·04 |

the cumulative birth-rate is therefore equivalent to the migration rate at the top of the crypt: this is about 1·8 cell positions per hour.

If we know the speed at which cells are moving at each cell position, then a simple calculation allows an estimate of the time taken to traverse that position; this is the reciprocal of the velocity. Adding the times taken over various portions of the crypt gives their respective transit times. For example, the time taken for a cell to migrate from cell position 4 to cell position 29 is about 32 hours. If the mean cell cycle time is about 11 hours, then some three transit divisions occur between cell position 4 to 29. Note that above position 29 only maturing cells, showing no mitotic activity, are evident.

We have chosen as the upper limit of the proliferative compartment the uppermost cell showing proliferative activity. Inspection of Fig. 7.2 shows that proliferative activity decreases comparatively slowly as we move up the crypt. This phenomenon has been explained in terms of a linear decrease in the probability of *proliferative* cells being produced as we move up the crypt. We can use the trailing edge of the labelling index curve shown in Fig. 7.3 as a reference point, and thus show that it takes about 11 hours for this point to reach the top of the crypt. Allowing time for labelled cells to complete mitosis, we find that it takes about 6 hours for a non-proliferative cell to pass through the maturation sequence after completing a final mitosis. The transit time through the functional compartment (villus) has been estimated at about 48 hours.

The turnover time for the dividing compartment (where by definition all cells are proliferative) would be expected to be close to $t_C$, or 11 hours. However, this is an FLM estimate; preference is therefore given to cell positions with most mitosis, i.e. the higher cell positions (Fig. 7.5). Because $t_C$ values are longer in basal positions the turnover time could be longer. In the maturation and functional compartments the turnover time should equal the transit time, i.e. about 6 hours and 48 hours respectively. Experimental estimates of turnover time have been shown to lie between 20 hours and 30 hours in the crypt, and up to 40 hours for the villous epithelium.

Table 7.2 Kinetic parameters in the rat jejunal mucosa.

| Parameter | Proliferative (dividing) | Maturation | Functional |
|---|---|---|---|
| Size | 400 | 200 | 3000* |
| Cell cycle time (hours) | 11·3 | – | – |
| Transit divisions (m) | 3 | – | – |
| Birth rate (cells/hour) | 35 | – | – |
| Input $k_{in}$ (cells/hour) | – | 35 | 35* |
| Output $k_{out}$ (cells/hour) | 35 | 35 | 35 |
| Transit time (hours) | 32 | 6 | 48 |
| Turnover time (hours) | 11·3 | 6 | 48 |

* The situation is complicated by the fact that about 20 crypts feed one villus.

So much for a normal cell population in a steady dynamic state. We turn now to consider expanding populations of abnormal cells. The reader should note the relative paucity of experimental results in the human case, due to the constraints of medical ethics.

## 7.2  Experimental study of a Balb/c mouse sarcoma

The study of the indigenous mouse sarcoma, our second worked example, illustrates well the value of combining a number of techniques to obtain a better understanding of the cell population under study. It also highlights two problems to which insufficient attention is paid. The first is the degree of kinetic heterogeneity within a population. We have seen that in 'normal' tissues it may be possible to discern distinct 'compartments' within a cell population, and that cells in different compartments may have different kinetic properties. In tumours, evidence of such organization is usually obscure or absent, depending on the degree of differentiation. Within a given locus the tumour may comprise a mixture of stem cells, proliferating non-stem cells, $G_0$ resting cells, differentiating end state cells, and dying cells, (refer to Fig. 2.2). These cells are not easy to tell apart, but their existence cannot be ignored. The second consideration highlighted in this and in a number of other studies to be found in the literature, is that the kinetics of a cell population may change with time. We shall see that the growth of the mouse sarcoma slows down as the tumour ages. There are other ways in which the kinetics of a cell population, even a normal one, may vary with time. For example, diurnal variations have been demonstrated in a number of tissues; this may make the interpretation of kinetic data more difficult. And, of course, both growing and steady state populations will show changes in kinetic parameters after perturbation by factors such as radiation or cytostatic drugs.

The tumour studied arose spontaneously in an indigenous castrate male Balb/c mouse and has been successfully transplanted through 150 passages since the original transplantation. The experiments to be described were carried out between the 95th and 110th passage. The tumours were transplanted sub-cutaneously. Animals were killed and their tumours removed and weighed at intervals up to 21 days after transplantation. From these weights the growth curve shown in Fig. 7.6 was constructed on a semi-logarithmic grid. It is apparent

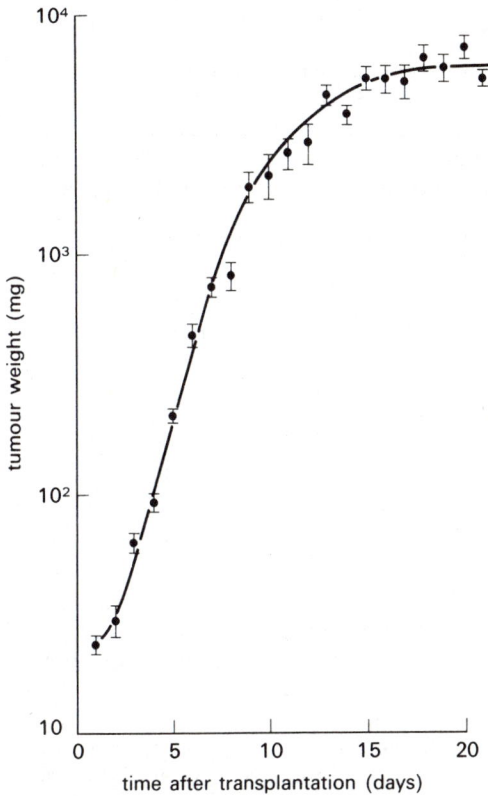

**Fig. 7.6** The growth curve for the Balb/c sarcoma. The experimental points indicate mean and SE, but note that the plot is semi-logarithmic. This tumour demonstrates a typical Gompertzian form of growth.

that tumour weight increases linearly up to day 6 or 7; over this period, therefore, the tumour grows exponentially. However, after 6–7 days the growth curve flattens markedly. Kinetic investigations were carried out after 7, 14, and 21 days to discover which parameters were changing and thus causing the observed slowing in growth rate. We will describe the results for the 7-day tumours first and then go on to compare these with the results obtained for the 14- and 21-day tumours.

Before we proceed, some passing comments on the mode of growth and the sampling procedure are apposite. The tumours grew as spherical or ovoid sub-cutaneous masses. Necrosis appeared to be confined to the centre and was present even in quite small tumours; the necrotic centre increased in size as the tumours became larger. In order to study how kinetic parameters varied with anatomical site, histological sections were prepared along the major axis of the tumour, so that the necrotic centre could be visualized in relationship to the encapsulated periphery. The viable tumour was then divided into three approximately equal zones. These zones are not intended to be compartments; it is not feasible to partition tumours in this way.

### 7.2.1 *Kinetics of the 7-day tumours*

(a) $^3H$-*thymidine studies*    Animals were pulse labelled with an injection of $^3$H-TdR at a dose of 1 $\mu$C/g of body weight $^3$H-TdR injection. Six animals were killed one hour after injection and their tumours were removed to enable labelling and mitotic indices to be determined. In Table 7.3 we see the indices for the whole tumour and each zone; it is evident that the indices are lower in zone 3.

**Table 7.3** Mitotic and labelling indices in the 7-day tumours (Balb/c sarcoma).

|  | $I_S$ | $I_M$ |
|---|---|---|
| Whole tumour | 0·28 ± 0·15 | 0·024 ± 0·0024 |
| Zone 1 | 0·33 ± 0·014 | 0·027 ± 0·0011 |
| Zone 2 | 0·35 ± 0·013 | 0·027 ± 0·0014 |
| Zone 3 | 0·12 ± 0·013 | 0·014 ± 0·0015 |

In all animals the fraction of labelled mitoses was determined in each tumour by counting at least 100 metaphases and anaphases per zone. The resulting FLM curves are shown in Fig. 7.7. Figure 7.7a is the curve for the whole tumour and has two well defined peaks. The lines were fitted to the experimental points using a computer programme, so that means and standard errors could be calculated for the cycle and each of its phases. In Table 7.4 we have quoted the means for each phase with their standard errors (SE). The validity of computer SEs is still not fully accepted. They are considered by many to be too small, possibly by a factor of about 2. We shall not use them in any significance testing but they are useful in estimating the spread in cell cycle times in the cell population for which the curve was constructed. The mean cycle time of 13·4 hours for the whole tumour is short even for rapidly growing mouse tumours. The standard error is small as would be expected for a curve with such a well defined second peak.

The curves for the three zones (Figs. 7.7b, c, and d) are similar and as we can see in Table 7.4 there is no significant difference between the cell cycle time ($t_C$) in the various anatomical sites at day 7. This is in agreement with a number of investigations on other solid tumours; in general $t_C$ has been found to remain constant even when other parameters have shown marked changes.

(b) *Stathmokinetic data*    Again at days 7, 14, and 21, animals were given vincristine (1 mg/kg body weight) intraperitoneally at 9.00 a.m. Animals were then

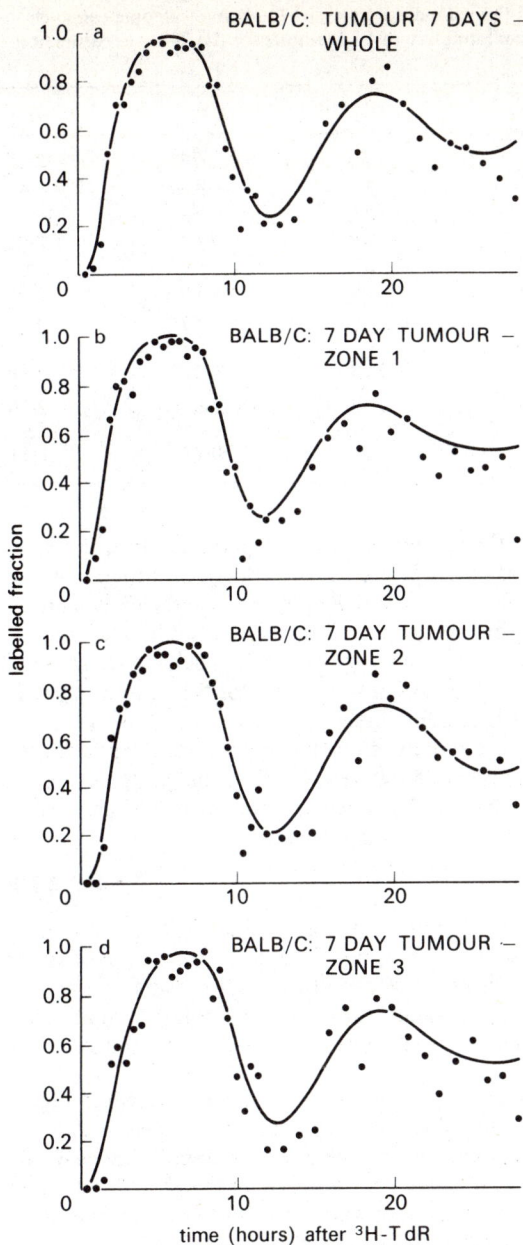

**Fig. 7.7** The fraction of labelled mitoses curve for the whole tumour (*a*), and the three zones described in the text (*b, c,* and *d*).

**Table 7.4** The phase durations, and their standard errors, of the Balb/c sarcoma cell cycle, as determined by the FLM technique. Summary of FLM results for the 7-day experimental tumour of the mouse.

| | $t_C$ | $t_{G1}$ | $t_S$ | $t_{G2}$ |
|---|---|---|---|---|
| Whole tumour | 13.41 ± 0·29 | 3·30 ± 0·29 | 7·74 ± 0·24 | 2·36 ± 0·11 |
| Zone 1 | 13·08 ± 0·32 | 3·43 ± 0·30 | 7·64 ± 0·20 | 2·01 ± 0·09 |
| Zone 2 | 13·82 ± 0·27 | 3·85 ± 0·28 | 7·69 ± 0·21 | 2·28 ± 0·10 |
| Zone 3 | 13·07 ± 0·30 | 2·67 ± 0·28 | 7·63 ± 0·23 | 2·78 ± 0·11 |

killed at 15 minute intervals and the mitotic index was determined in each tumour by scanning at least 3000 nuclei per zone. The mitotic accumulation curves for zones 1, 2, and 3 are plotted in Fig. 7.8 *a, b,* and *c,* and clearly a good linear graph was obtained in all three zones. The lines were fitted by regression analysis and the 95 per cent confidence limits are shown for each curve. There is a slight delay before the collection lines begin to rise and for this reason the initial readings were excluded in fitting the line.

From the lines in Fig. 7.8 we can obtain cell birth-rates for each zone and, as Table 7.5 shows, the birth-rate is progressively reduced from the periphery to the centre of the tumour. It may be noted at this point that birth-rate can also be calculated from data obtained in the labelling studies:

$$k_B = \ln(1 + I_p)/t_C. \qquad (7.4)(cf.\ 2.10)$$

These results calculated in this way do not differ significantly from the vincristine data and so for the sake of brevity we do not show them here. By combining the vincristine birth-rates for the three zones we get some idea of the overall birth-rates for the whole tumour. This value is also given in Table 7.5, as are the calculated apparent cell cycle times for each zone and the whole tumour.

(c) *Assessment of growth*    The growth curve for this tumour is shown in Fig. 7.6. Even by day 7 the curve has probably begun to depart from the initial straight line and so, to find the growth rate, tangents were drawn to the curve at days 7, 14, and 21. The slope of the tangent is a measure of the growth rate ($k_G$) and from this, by assuming exponential conditions for a short period where the tangent touches the curve, we can calculate the doubling time, thus:

$$t_D = \ln 2/k_G. \qquad (7.5)(cf.\ 2.14)$$

It is also possible to obtain a good fit to this data using a Gompertz function. From the equation to the line so obtained the growth rate at any time can be calculated directly. However, for brevity, we will give only the results obtained

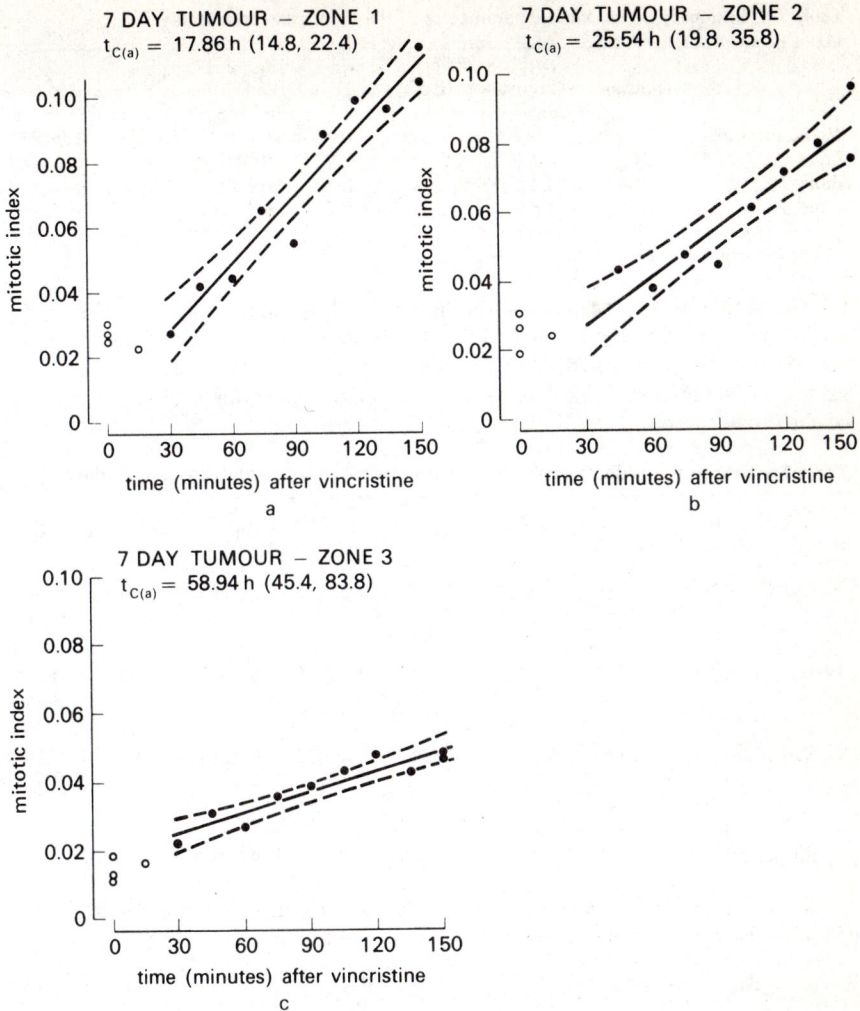

**Fig. 7.8** The mitotic index accumulation lines for the three zones of the Balb/c sarcoma, after vincristine injection. Zone 1, (*a*); zone 2, (*b*); zone 3, (*c*).

from the tangent method, since the two methods are in good agreement and the choice will not affect our basic conclusions.

At day 7 an estimate of 0·53 mg per mg per day was obtained for $k_G$ and from this the doubling time was calculated to be 1·4 days (33·6 hours).

(d) *Growth fraction and cell loss*     As we saw in Chapter 6 (p. 56) the growth fraction can be calculated in two ways from the data available. We can calculate the theoretical labelling index, assuming exponential growth, from equation **(2.6)** and by comparing this with the experimental labelling index, obtain the growth fraction. Alternatively, knowing $t_C$ from the FLM curve and $t_{C(a)}$ from

**Table 7.5** Summary of the kinetic parameters of the Balb/c sarcoma at day 7.

| | $t_C$ (hours) | $t_{C(a)}$ (hours) | $t_D$ (days) | $I_P$ | $k_B^*$ | $k_G^*$ | $k_L^*$ |
|---|---|---|---|---|---|---|---|
| Whole tumour | 13·4 | 26·7 | 1·32 | 0·50 | 0·037 | 0·0219 | 0·0155 |
| Zone 1 | 13·1 | 17·9 | – | 0·73 | 0·056 | – | – |
| Zone 2 | 13·8 | 25·5 | – | 0·54 | 0·039 | – | – |
| Zone 3 | 13·1 | 58·9 | – | 0·22 | 0·017 | – | – |

*Cells per cell per hour.

the vincristine data, we can calculate the ratio $t_C/t_{C(a)}$ and thus obtain the growth fraction. For each zone and for the tumour as a whole, we therefore have two estimates of the growth fraction which are in reasonable agreement. These values are included in Table 7.6, which summarizes the major kinetic parameters of the 7-day tumour.

**Table 7.6** Summary of kinetic parameters of the Balb/c sarcoma at 7, 14, and 21 days.

| | $t_C$ (hours) | $t_{C(a)}$ (hours) | $t_D$ (days) | $k_B^*$ | $k_G^*$ | $k_L^*$ | $\phi$ | $I_P$ |
|---|---|---|---|---|---|---|---|---|
| 7 Days | 13·41 ± 0·29 | 26·7 | 1·32 | 0·037 | 0·0219 | 0·015 | 0·15–0·5 | 0·50 † 0·45‡ |
| 14 Days | 12·82 ± 0·41 | 27·9 | 6·94 | 0·036 | 0·00416 | 0·032 | ~0·8 | 0·46† 0·38‡ |
| 21 Days | 13·11 ± 0·39 | 37·7 | 99·5 | 0·026 | 0·00029 | 0·026 | ~0·9 | 0·35 † 0·41‡ |

* Cells per cell per hour.   † Calculated from $t_C/t_{C(a)}$.   ‡ Calculated from $I_{S(ob)}/I_{S(th)}$.

Knowing the birth rate ($k_B$), from the vincristine study, and the growth rate ($k_G$) from our growth curve, we can calculate cell loss rate ($k_L$) from

$$k_L = k_B - k_G. \qquad \text{(7.6)(cf. 2.13)}$$

We can establish the relative contributions of cell birth and cell loss by calculating the cell loss factor ($\phi$) from

$$\phi = k_L/k_B = 1 - (t_{C(a)}/t_D). \qquad \text{(7.7)(cf. 2.15)}$$

The values we obtain for $k_L$ and $\phi$ are given in Table 7.6.

Two points are worth emphasizing at this stage, namely:

(1) in general, vincristine and labelling data are in good agreement;
(2) the cell cycle time remained constant in the different zones despite changes in other parameters.

### 7.2.2 The 14- and 21-day tumours

A problem posed by the 14- and 21-day tumours was the poor penetration of $^3$H-label into zone 3. This prevented the construction of separate FLM curves for the three separate zones; only one curve was constructed based on zone 1 plus zone 2. Mitotic arrest experiments were unaffected in the inner zone, and we may speculate that some factor, or complex of factors, is interfering with $^3$H-Tdr uptake. In the zones concerned there is a good deal of necrosis, with breakdown of nuclei and the consequent release of DNA catabolites. It is possible that the concentration of metabolites is sufficient to inhibit uptake of $^3$H-Tdr. This hypothesis would account for irregular or even complete absence of labelling in the presence of measurable mitotic activity. Proliferative indices are lower in the central zone and the vincristine data show that cell production is reduced. Thus the subsequent discussion is concentrated on comparing data for the whole tumours at the different time points.

(a) *FLM data*    The curve for day 14 is shown in Fig. 7.9 and for day 21 in Fig. 7.10. At both times two well-defined peaks can be seen, but on both curves

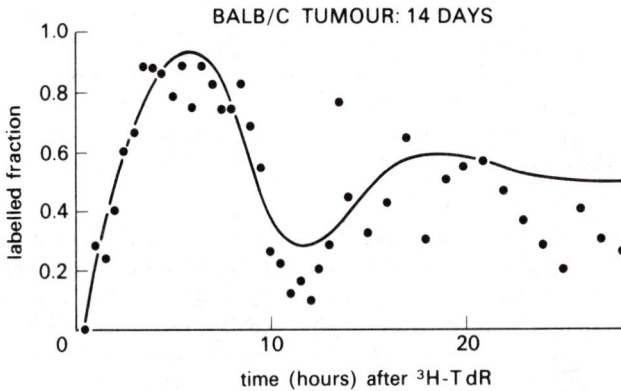

Fig. 7.9 The fraction of labelled mitoses curve for the Balb/c sarcoma at 14 days after transplantation.

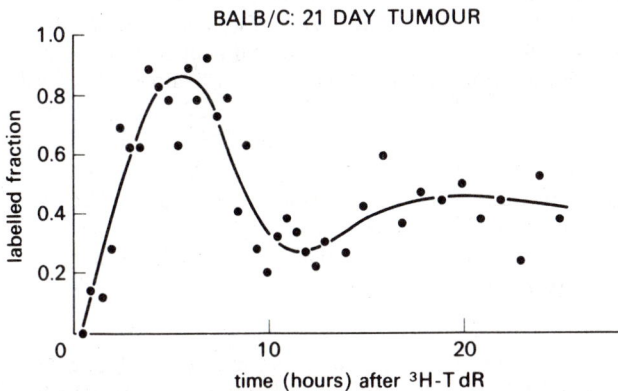

Fig. 7.10 The fraction of labelled mitoses curve for the Balb/c sarcoma at 21 days after transplantation.

the height of the peaks is reduced and damping of the second peak is particularly noticeable compared with the 7-day curve. This damping is due to the greater variability of $t_C$ and its phases and this is reflected in the larger SEs calculated for $t_C$ at 14 and 21 days, (see Table 7.6). However, the mean values for $t_C$ at 7, 14, and 21 days are very similar. Thus we may conclude that the slowing in the growth rate of this tumour with age after transplantation is not due, even partially, to a prolongation of $t_C$.

(b) *Stathmokinetic data*   If we look again at Table 7.6, we will see the birth-rates for the whole tumour at 7, 14, and 21 days. These results, 0·037, 0·036, and 0.026 cells per cell per hour at 7, 14, and 21 days respectively, suggest that the rate of cell production does not decrease much over the period of tumour evolution; certainly not over the first 14 days. There is a small decrease in cell production rate by 21 days. This is reflected in the $t_{C(a)}$ values calculated from the vincristine data: 26·7 hours at 7 days, 27·9 hours at 14 days, and 37·7 hours at 21 days.

(c) *Growth measurements*   At 14 days the growth rate had slowed considerably compared with day 7; the tangent method gave a value for $k_G$ of 0·0999 mg per mg per day. At day 21, $k_G$ was calculated to be much lower at 0·00696 mg per mg per day. This is not surprising when we note how flat the growth curve has become at this point. The corresponding doubling times are: 14 days − 6·9 days; and at 21 days − 99·5 days.

(d) *Growth fraction and cell loss*   Again growth fraction can be calculated by the two methods described previously and the two values for each time point are given in Table 7.6. At 14 days the two estimates are in good agreement and are of very similar magnitude to those calculated for the 7-day tumours. At 21 days the two estimates do not agree so well; the problem of uneven labelling is such that it might be better to accept the value calculated from the ratio $t_C/t_{C(a)}$ as more reliable. There would seem to be a fall in growth fraction by 21 days and thus we may conclude that a decrease in growth fraction plays a small part in the diminution in growth rate observed during development of the tumour.

It is difficult to calculate the cell loss rate accurately. As we saw earlier, it is possible to calculate birth-rates from the labelling data as well as from the vincristine data: the choice of method does not affect the main conclusions. However, rather than quote a single figure we have given a range of values calculated by the different methods for $\phi$ at each time point (Table 7.6). At 7 days values for $\phi$ lie mainly between 0·15 and 0·5, increasing to about 0·8 at 14 days. At 21 days most values are about 0·90, and several are above 0·98. Thus at 7 days, for every 100 cells born about 50 are lost, at 14 days about 80 are lost, and at 21 days for every 100 cells as many as 98 are lost.

We may conclude that the main reason for slowing of growth rate during tumour evolution is the increased rate of cell loss.

### 7.2.3 *Microdensitometry*
Touch preparations were made from freshly resected tumours at 7, 14, and 18 days after implantation and stained by the Feulgen technique. Estimations of the DNA content of individual nuclei were obtained by microdensitometric measurement of the light absorption at a wavelength of 550 nm, using an in-

tegrating microdensitometer. The results were expressed as described in Fig. 7.11. Mouse sperm nuclei were assumed to represent haploidy and were therefore used as the reference value for tumour nuclei.

In Fig. 7.11 we see the DNA profiles of the tumour at 7, 14, and 18 days. At 7 days it is clear that the modal value is 2N; in other words, most cells contain the normal diploid amount of DNA. However, as the tumours age, the inter-nuclear variation becomes larger, indicating a progressive increase in polyploid and aneuploid nuclei. This is particularly noticeable at 18 days. The high degree of aneuploidy and polyploidy in the older tumours suggests either that some cells synthesize DNA but do not proceed to mitosis, or that grossly unequal mitotic division is occurring. With particular reference to the 18-day tumours, the highly polyploid and aneuploid nuclei might represent cells with limited proliferative potential (although some of these have been observed to undergo mitosis). These cells could be contributing to the lower growth fraction observed in the older tumours.

### 7.2.4 *Summary and comments on the mouse tumour studies*

Initially *tumour growth* is close to exponential; this is followed by a deceleration in growth rate and finally a period of apparent stasis. The growth curve can be described by a Gompertz function.

FLM measurements of *cell cycle time* showed no change in mean values, either with increasing age, or cell localization within the tumour. Our findings, and other published work, suggest that the cell cycle time may be a relatively constant parameter within any solid tumour. It is interesting to note that this does not seem to be the case with ascites tumours, in which considerable changes in the duration of the cycle have been found as the tumours age.

The *growth fraction* was found to vary with site in this tumour, being less in zone 3 than at the periphery. This explains the lower proliferative indices and cell production rate found in the inner zone. A small decrease in growth fraction was also found as the tumours aged.

Although the small decrease in growth fraction may contribute to the slowing in growth rate, the main reason for the growth retardation was an increase in the *cell loss rate*, with terminal $\phi$ values of the order of 0·98.

The reason for these kinetic changes is not clear; there may be specific growth control mechanisms at work. However, it is possible that changes in growth fraction and cell loss are a direct function of tumour size. As the tumours get larger, restrictions on the supply of oxygen and nutrients may become more severe causing more cells to cease proliferation and die, particularly in the centre of the tumour.

This study illustrates some of the practical problems which may be encountered in a kinetic investigation of a real cell population; for example, the uneven labelling found in larger tumours and the phenomenon of kinetic heterogeneity. However, by combining a number of techniques we have gained a good understanding of the kinetic properties of this tumour cell population.

## 7.3 Human tumour kinetics

Much of the impetus for kinetic studies has come from a desire, ultimately, to improve the treatment of cancer in humans. Further, much of the financial sup-

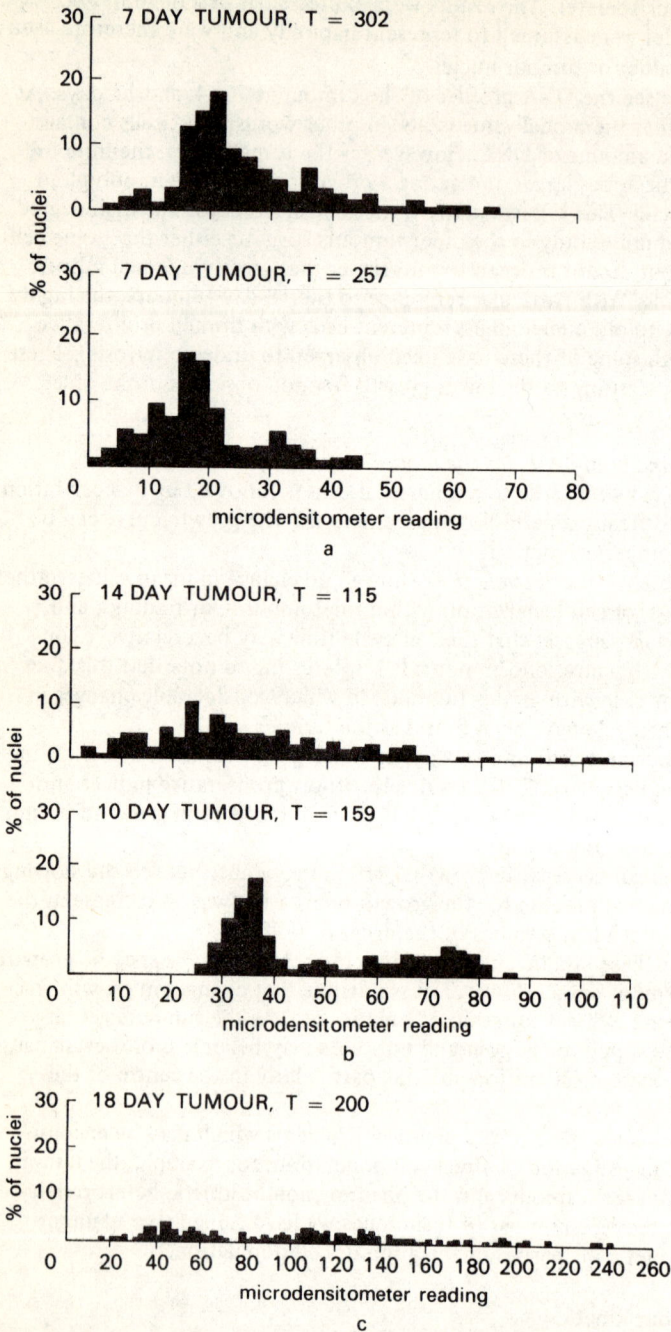

Fig. 7.11 DNA histograms at various times after transplantation of the Balb/c sarcoma. Note the wide spread at 18 days, due to aneuploid and polyploid nuclei.

port for such investigations has come from cancer research funds. For this reason, despite the obvious ethical and practical problems, attempts have been made to study the kinetics of human tumours both *in vivo* and *in vitro*. *In vivo* studies have of necessity been limited and such detailed investigations as that described for the mouse tumour have not been carried out. Nevertheless it is interesting to compare human tumours with transplantable animal tumours. Some of the limitations of animal systems as models for human cancer can then be seen.

Take, for example, rectal carcinoma, a common tumour of middle-aged or elderly patients. We carried out an *in vivo* stathmokinetic study, using vincristine, and from this calculated cell birth-rates and apparent cell cycle times. For a group of 19 large bowel carcinomas, a mean $k_B$ of 0·6 cells/100 cells per hour (SE 0·05) and a mean $t_{C(a)}$ of 170 hours were found. We were unable, for ethical reasons, to carry out *in vivo* labelling studies. However, in the literature we found a report of an FLM study carried out on a similar group of tumours, in which $t_C$ was calculated to be between 24 and 48 hours. Also in the literature we found a mean estimate for the doubling time of a group of 20 large bowel carcinomas. This value was 620 days (or 15 000 hours).

The reader will recall that growth fraction is given by the ratio $t_C/t_{C(a)}$, and from the equation $1 - (t_{C(a)}/t_D)$ we can calculate the cell loss factor. These estimates are inevitably crude since the values for $t_C$, $t_{C(a)}$ and $t_D$ were obtained by different workers on different cases of rectal carcinoma. Further, in contrast to the animal investigation, it was not possible to study tumours at different stages of their development; nor was it possible to see how kinetic parameters varied with location in the tumour. Despite the crudeness of the estimates, several points of interest emerge (see Fig. 7.12). The cell cycle time is longer in human

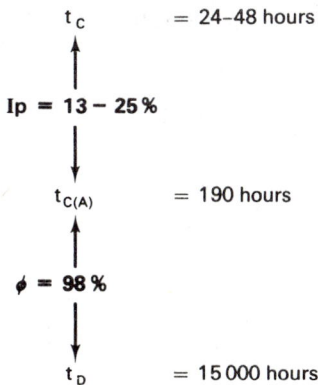

$t_C$ = 24–48 hours

Ip = 13 – 25 %

$t_{C(A)}$ = 190 hours

$\phi$ = 98 %

$t_D$ = 15 000 hours

**Fig. 7.12** The relationship between the various kinetic parameters as measured in human rectal carcinoma.

rectal carcinoma than in the mouse tumour, the growth fraction is considerably lower, and cell loss is high. It is a general phenomenon that human tumours grow more slowly than transplanted animal tumours. This is an important factor if animal tumours are used as models for evaluating the possible usefulness of therapeutic schedules for the treatment of human tumours, especially when these schedules involve the use of 'cell cycle specific' agents. These are compounds which (like X-irradiation) act preferentially on proliferating cells and often on cells in a particular phase of the cell cycle. In rectal carcinomas and in other types of human cancer, the existence of a large 'non-growth fraction' of presumably resistant cells may well explain why human tumours generally do not show as good a response to cycle-specific therapy as do many animal tumours.

However, animal and human tumours may not necessarily be as different kinetically as these two examples would suggest. Firstly, there is good evidence that spontaneous animal tumours grow more slowly than tumours which have been frequently transplanted. Thus spontaneous animal tumours may be a little more like spontaneous human tumours. Secondly, the human tumours encountered clinically are well along their growth curves, i.e. they are quite 'old' tumours. As we saw above, the growth rate of the mouse tumour slowed dramatically as the tumours aged, and values for cell loss factor similar to those of rectal tumours were found.

The possible relevance of kinetic data to the clinical treatment of malignant disease is a complex topic outside the scope of this short book. To date there is not much evidence that tumour therapy has been made more effective by the use of such data. However, kinetics is a relatively new field of study and one cannot discount the possibility that future kinetic studies may be of clinical use. Whether they are or not, the study of cell kinetics of both normal and diseased tissue is certainly of considerable scientific interest.

## References

BOOTH, C. C. (1968) Enteropoiesis: structural and functional relationships of the enterocyte. *Postgrad. Med. J.*, **44**, 12.

GILBERT, C. W. (1972) The labelled mitoses curve and the estimation of the parameters of the cell cycle. *Cell Tiss. Kinet.*, **5**, 53.

# Glossary of cytokinetic terms

| | |
|---|---|
| Aneuploid | A state in which the amount of chromatin in a nucleus does not equal an integral multiple of the haploid or diploid amount (see Feulgen; polyploid). |
| Antitemplates | Hypothetical substances which are supposed to depress or inhibit cell proliferation (see templates). |
| Autoradiography (ARG) | A technique for detecting the presence of a radioactive source. In cell population kinetics the source is usually tritium ($^3$H), sometimes carbon ($^{14}$C), both of which may be incorporated into DNA (q.v.). Their presence in so-called 'labelled cells' is detected by placing a layer of photographic emulsion over the cell preparation (section, imprint, or smear) in which silver 'grains' are deposited by $\beta$-particles emitted from the radioactive atom. |
| Carcinoma | A malignant tumour which arises from epithelial tissues (cp Sarcoma). |
| Cell cycle | A sequence of events in the life history of proliferating cells, usually regarded as beginning with mitosis and ending with the onset of the next mitosis. |
| Chalone | A substance or substances believed to have a specific inhibitory action on cell proliferation. Chalones are tissue specific but not species specific. Evidently they can act on different phases of the cell cycle; for example in late $G_1$ and in $G_2$. The physiological role of these substances is, as yet, uncertain. |
| Colcemid | An analogue of colchicine, less toxic, with the property of arresting dividing cells in metaphase. This compound (no longer commercially available) has several defects as a stathmokinetic agent (q.v.). It has been largely superceded by the vinca alkaloids vincristine and vinblastine. |
| Colchicine | An alkaloid derived from the autumn crocus, which arrests cells in metaphase. |
| Collection function | In stathmokinetic experiments where the age distribution of the proliferating cells is exponential, it is usual to plot $\log_{10} (1 + I_m)$ as the dependent variable. This is termed the collection function. |

| | |
|---|---|
| Continuous labelling | A technique by which tritiated thymidine is made continuously available to the proliferating cells; *in vivo* by infusion or serial injection, *in vitro* by continued presence in the culture medium. |
| Cytotoxic | Compounds which are capable of killing proliferating cells. Cytotoxic chemicals are usually divided into cycle specific (for example cytosine arabinoside) which act selectively at a period in the cell cycle (in this case DNA synthesis), and non-cycle specific agents, which characteristically act at several distinct periods in the cycle. |
| Diploid | The usual amount of chromatin (number of chromosomes) in a somatic cell prior to the initiation of DNA synthesis. |
| DNA | Deoxyribonucleic acid, a double chain composed of nucleotides and a ribose phosphate backbone, disposed in a helical formation. This is the genetic material of the cell. The sequences of nucleotides in the chain are essential in coding the sequence of bases in ribonucleic acid, and hence the order of amino acids in proteins synthesized at the ribosomes. |
| DNA synthesis | The process by which the nucleus replicates its complement of DNA. The replication of DNA chains is semiconservative; each chain serves as a template for a new one. During DNA synthesis, certain metabolites, such as thymidine triphosphate, are selectively incorporated into DNA; if the thymidine is labelled, with tritium or $^{14}C$, then such labelled nuclei are detectable by autoradiography. |
| Double labelling | An autoradiographic method which employs the principle of labelling a cohort of cells first with one specific label, such as $^{14}C$-thymidine, and after a measured period, a second such as $^3H$-thymidine. By identifying the cells labelled with $^{14}C$-labelled nuclei, $^3H$-labelled nuclei, and $^{14}C$- and $^3H$-labelled nuclei, the rates of entry into and out of DNA synthesis can be calculated. |
| Enteroblast | A term applied to the proliferating cells found in the crypts of Lieberkühn of the small bowel, which differentiates to form the functional enterocyte (q.v.). |
| Enterocyte | The functional cell of the small bowel, a layer of which covers the villi. Enterocytes are responsible for the absorption of nutrients from the bowel lumen. |
| Epidermis | The epithelial tissue which covers the surface of the body, i.e. the epithelial component of the skin. It is composed of stratified (layered) epithelium. Proliferation is confined to the basal layers, and cells migrate towards the surface to be shed as anuclear keratinized forms. |
| Epithelium | Tissue which covers or lines surfaces, such as the skin and the gastrointestinal tract. |

| | |
|---|---|
| Exponential | An increasing function, written $y = e^x$ or $y = \exp(x)$. Convenient as an approximate growth equation (but see Gompertz). Also used of age distribution diagrams in which, because the population is expanding, the incidence of 'young' cells is highest (cf. rectangular distribution). |
| Feulgen | A stain which depends on the combination of *Schiff's reagent* with DNA; the amount of stain incorporated is proportional to the amount of DNA present. Consequently by measuring the relative amount of stain by micro-densitometry, the relative amount of DNA in a nucleus can be estimated. Nuclei can then be classified according to their DNA content relative to the diploid amount. |
| FLM | The abbreviation for 'fraction of labelled mitoses' technique, sometimes called percentage labelled mitoses (PLM) technique, for measuring the cell cycle time and its component phase durations. Cells in DNA synthesis are labelled with tritiated thymidine, and their progress around the cell cycle is followed by plotting consecutive waves of labelled mitoses against time. |
| Germ cells | These are the spermatozoa and oogonia, i.e. the specific cells of the testis and ovary respectively. |
| Goblet cells | These cells are found characteristically in the small and large intestinal mucosae, and are of characteristic 'goblet' shape. Current thinking derives these cells from undifferentiated proliferating crypt cells. |
| Gompertz | A mathematical function used to fit certain growth curves of a sigmoid shape, such as that of tumours, regenerating liver nodules, and many organs during organogenesis. In brief the Gompertz process may be regarded as compounded of exponential growth and simultaneous exponentially increasing retardation of growth. |
| Grain count | The number of silver grains, in autoradiographs, which are located above a labelled nucleus. Thus at mitosis, if all cells divide, the *mean grain count* should be halved. |
| Haploid | Cells which contain half the complement of DNA of that found in diploid somatic cells; during reduction division, or meiosis, cells in the male testis, and in the female ovary acquire the haploid amount of DNA. |
| Isoproterenol | A drug which mimics the effect of stimulation of the sympathetic nervous system. It has the property of stimulating a proliferative response in the rodent salivary gland. |
| Labelling | In the context of this book, incorporation of $^3$H-TdR or $^{14}$C-thymidine during DNA synthesis, which can later be demonstrated by autoradiography or by scintillation counting. |
| Large bowel carcinoma | See rectal carcinoma. |

| | |
|---|---|
| Lymphocytes | Cells produced by lymph nodes, thymus, and spleen, which are responsible for both humoral and cellular immune reactions. |
| Maturation | The process by which cells progress from an undifferentiated, usually proliferative state, to a differentiated, non-proliferative state in which the cell is capable of its specific function. Thus the maturation compartment is one in which the process of maturation occurs. |
| Metaphase | One phase of mitosis, occurring after prophase and before anaphase and telophase, in which chromosomes became arranged at the centre of the *spindle*, to the poles of which they are drawn during anaphase and telophase. Agents which arrest cells in metaphase are spindle poisons. |
| Microdensitometry | A technique used (in the present context) to measure the intensity of Feulgen staining in individual nuclei, and hence to assess the amount of nuclear DNA. |
| Necrosis | Death of cells, caused for example by lack of oxygen and nutrients, in the centre of a solid tumour. This is one of the mechanisms of cell loss. |
| Neurone | A species of cell in the central nervous system, responsible for the reception and transmission of nervous impulses. These cells do not normally divide postnatally. |
| Nucleotides | The building blocks of the nucleic acids. They are composed of a purine or a pyrimidine base attached to a carbohydrate (ribose or deoxyribose). If a nucleotide is labelled with a radioisotope which is later incorporated into DNA or RNA, then the synthesizing cell can be identified. |
| Ovary | The organ in the female which is responsible for the production of eggs, or *oogonia.* |
| Paneth cells | These are found at the base of the small intestinal crypt and contain numerous cytoplasmic granules. They apparently originate from undifferentiated, basally situated crypt cells, but as yet their function is undefined. |
| Polyploid | A cell containing a multiple of the normal diploid amount of DNA, e.g. tetraploid, octaploid, etc. |
| Production | A term applied to the rate at which cells are produced with respect to the population as a whole, proliferative, and non-proliferative. |
| Proliferation | Literally 'the bearing of offspring'. The term is used of cells actually in cycle, and refers usually to the rate at which they generate daughter cells (cf. cell *production*). |
| Radiotherapy | A method of treating cancer by irradiation, usually with gamma rays; these rays act selectively on proliferating cells. |
| Rectal carcinoma, (large bowel carcinoma) | A 'malignant' (i.e. cancerous) growth which originates in the mucosa of the large bowel or rectum. |

| | |
|---|---|
| Rectangular age distribution | An equiprobable distribution, i.e. the probability that a cell, picked at random from a population, has a given age (in terms of cell cycle time) is uniform over the whole population. As such it can be represented by a rectangle (cf. exponential age distribution). |
| Reticulocyte | A final stage in the maturation of red blood cells which is usually confined to the bone marrow. |
| RNA | *Ribonucleic acid*, a macromolecule responsible for the transfer of information from the DNA chain to the *ribosomes* in the cytoplasm, where protein synthesis occurs. |
| Sarcoma | A malignant tumour derived from connective tissues (cp Carcinoma). |
| Scintillation counting | A technique for determining the absolute or relative amount of a radioactive precursor (e.g. $^3$H-TdR) incorporated into a tissue. It is usually expressed as counts per minute per $\mu$g DNA. |
| Smooth muscle | Muscle which is not under voluntary control, but is controlled by the autonomic nervous system. It is found in internal organs such as the gut wall, and is widely distributed in arteries and veins. |
| Squamous epithelium | A covering of flattened cells which covers surfaces. Stratified squamous epithelium is found covering the skin and lining the oesophagus. |
| Stathmokinetic | An epithet given to techniques which depend upon the arrest of cycling cells, usually in metaphase, and then measuring the rate of accumulation of arrested metaphases. Suitable agents for this technique include vincristine, vinblastine, and colchicine. |
| Steady state | A kinetic situation whereby for every cell produced one leaves the compartment or population. It should be appreciated that although over all a population may be in steady state, the age distribution of the proliferating cells may be exponential (q.v.). |
| Stem cell | A cell which is capable of continued self-renewal and which also usually supplies cells to a proliferative or maturation compartment. |
| Template | A hypothetical substance which is supposed to stimulate, or mediate, the formation of copies of itself. This process is supposed to be inhibited by 'antitemplates' (q.v.). |
| Testis | The male sex organ where the germ cells (spermatozoa) are produced. |
| Villus | The finger-like projection from the mucosa of the small intestine. |
| Vinblastine | An alkaloid prepared from the periwinkle plant, *Vinca rosa*. It is used in the chemotherapy of cancer. Its property of arresting cells in metaphase is exploited as a research procedure (see Stathmokinetic). |
| Vincristine | Like vinblastine this is also prepared from the periwinkle plant. It is probably the best available metaphase arrest agent. |

# Suggestions for further reading

Baserga, R. (Ed.) (1971) *The Cell and Cancer*. Marcel Dekker, Inc., New York.

Busch, H. (Ed.) (1971) Studies on tumor cell population kinetics. In: *Methods in Cancer Research* **6**, 4–93. Academic Press, New York.

Cleaver, J. E. (1967) *Thymidine Metabolism and Cell Kinetics*. North-Holland, Amsterdam.

Goss, R. J. (Ed.) (1972) *Regulation of Organ and Tissue Growth*. Academic Press, New York.

Lala, P. K. (1972) Age-specific changes in the proliferation of Ehrlich ascites tumor cells grown as solid tumors. *Cancer Res.* **32**, 628–36.

Machin, D. (1976) *Biomathematics: An Introduction*. The Macmillan Press Ltd, London.

Mitchison, J. M. (1971) *The Biology of the Cell Cycle*. Cambridge University Press, London.

National Cancer Institute Monograph 38 (1973) *Chalones: Concepts and Current Researches*.

Snedecor, G. W. and Cochram, W. G. (1967) *Statistical Methods, 6th Edition*. The Iowa State University Press, Ames, Iowa.

Symington, T. and Carter, R. L. (Eds.) (1976) *Scientific Foundations of Oncology*. William Heinemann Medical Books Ltd, London

# Index